ENVIRONMENTAL PROTECTION IN SWEDISH FORESTRY

Environmental Protection in Swedish Forestry

KATARINA ECKERBERG
Department of Political Science
University of Umea, Sweden

Avebury

Aldershot · Brookfield USA · Hong Kong · Singapore · Sydney

Published by

Avebury

Gower Publishing Company Limited,
Gower House, Croft Road, Aldershot,
Hants. GU11 3HR, England

Gower Publishing Company
Old Post Road, Brookfield, Vermont 05036
USA

British Library Cataloguing in Publication Data
Eckerberg, Katarina, 1952-
Environmental protection in Swedish Forestry
1. Sweden. Forestry. Environmental aspects
I. Title
634.9'09485
ISBN 0-566-07052-9

Printed in Great Britain by
Billing & Sons Ltd, Worcester

Contents

List of figures and tables

Figures

Tables

Preface

This book is, in the main, built on my doctoral thesis in political science (1987) and my licentiate thesis in forest economics (1986). It is the result of six years of research carried out jointly at the Departments of Forest Economics and Political Science in Umeå, Sweden.

The combination of forestry and political science is highly relevant in an analysis of environmental considerations in forestry. It permits an examination of both the actual results of forest-environmental policy in terms familiar to the forestry audience, and of the policy implementation issues from a broader social and political perspective.

Gunnel Gustafsson, professor at the Department of Political Science, University of Umeå, was my doctoral advisor. Her comments on my work were extremely valuable. Stina Lindström deserves special thanks for her endurance in finalizing the manuscript.

The National Environmental Protection Board (NEPB) financed the research project. I was able to finish my doctoral studies through a grant from the Swedish University of Agricultural Sciences. The University and the Royal Swedish Academy of Agriculture and Forestry also funded my visit to the United States to compare U.S. and Swedish forest-environmental policy.

I wish to express my gratitude to the above mentioned, including many others who helped me finalize this book.

Katarina Eckerberg

1 Introduction

The problem: Environmental degradation in forestry

Environmental degradation is a serious problem resulting from large scale forest operations based on present technology. In Sweden a considerable proportion of plant and animal species are threatened: 20 per cent of the vascular plants; 20 per cent of the land-living molluscs; 30 per cent of mammals and birds; and 50 per cent of reptiles and amphibians. [1] For many of these species, forestry activities are main contributors to the danger of extinction.

Public concern for the preservation of flora and fauna has become a major political issue during the last two decades - primarily because of large clearcuts and mechanized harvest. [2] In addition, the demand for forest recreation is growing. Environmentalists have protested and mass media reports of environmental mistakes have been frequent.

In order to maintain a diversity of plants and animals their natural habitats have to be preserved. Plant and animal species have complicated ecological relationships which make the effects of exterminating any species uncertain. There are ethical reasons for preserving all natural species. A diversified ecosystem is part of our cultural heritage. In addition, there may be long term economic arguments for maintaining naturally diverse ecosystems. The immediate gain from plants and animals is, of course, their direct nutritive and material usefulness. Researchers continuously detect new fields of application for plant and animal species for human use. It is impossible to foresee their potential uses in the future.

Concerns for economic profitability make forest management in Sweden today favour high production of a few tree species - primarily conifers. Silvicultural treatments aim at creating homogenous forest stands, as similar as possible in terms of tree species, age, height, diameter and spacing between trees. Forest activities cover more of Sweden's landbase than ever before in history. Although vast forest areas were clearcut hundreds of years ago (in connection with mining and iron works and for cleaning pasture lands) these activities were concentrated to specific regions. Today modern forest machines are found almost everywhere. Draining of wet forest land and use of chemicals in forestry, such as herbicides and fertilizers, have become more common during the last two decades. [3] All the above measures promote the growth of trees, but at the same time contribute to the reduction of natural diversity.

There is political consensus in the Swedish Parliament as to the general goal to preserve the endangered plant and animal species and to maintain a diversified forest ecosystem. This political goal may be more symbolic than real, implying that politicians may never have intended to transform it into general forestry practice. I have assumed, however, that political consensus exists on a real level - that politicians are seriously aiming at the preservation of a diversified forest ecosystem as well as willing to make efforts to implement their goal. The problems are then: By what means is the goal to be achieved, and who is going to pay for it - the government or the forest owners? With the current Swedish forest-environmental policy, accepted by Socialists as well as non-Socialists, protection of natural habitats and recreational values in Swedish forests is achieved by a combination of large forest reserves, paid for by the government, and general forest land regulations. The general measures to protect the environment are to be achieved without economic compensation from the government. Eventual costs thus rest with the forest owners.

National parks and nature reserves (of which some land is donated by the forest companies) have been created primarily for nature conservation. [4] However, these reserved areas do not alone suffice as refuges for threatened plant and animal species, or for forest recreation purposes. Environmental values must also be protected on the remaining forest land if a diverse ecosystem should be maintained.

The amendment of environmental protection requirements in the Swedish Forestry Act of 1974 along with a penalty clause in 1979 reflect a policy change in Swedish forestry. A new paragraph - § 21 - (from 1979) stipulates that environmental considerations must be shown on all forest land concerning forest operations (including regeneration, spacing and thinning, harvesting and road-building).

The public debate on environmental issues remains somewhat

controversial. In the early 1980s, however, a change in debate occurred, involving a change of attitudes among both environmentalists and economic foresters [5]. The need for environmental protection within forest operations is thus nowadays an urgent aim of most land managers in the Swedish forest industry as expressed by a number of forest companies' recent colourful pamphlets on environmental considerations in their forest management. The Swedish Forest Workers' Labour Union also encourages its members to attend "study circles" [6] in environmental protection. [7]

Aim of the study

Even though spokesmen of economic forestry are influenced by the intensive debate over environmental issues, which have been on the political agenda ever since the 1960s, the questions of implementing protection measures remain. It is therefore interesting to examine what characterizes the implementation process and what factors determine the result of environmental protection.

The study seeks to describe and analyse the implementation of environmental protection in Swedish forestry, especially with clearcutting [8] operations. The first task is to analyse to what extent environmental features are actually protected in Swedish forestry and how different aspects of environmental protection, such as recreational values as well as protection of flora and fauna habitats are considered.

The next step is to discuss what factors determine the result of environmental protection. The investigation includes data on what actors are involved in the implementation process and how they promote or impede environmental protection. More precisely: Do the people who work in the woods actually pay attention to environmental interests? To what extent does the Forestry Act regulation concerning environmental protection affect their behaviour? Is there a discrepancy between policy goals and policy outcome? If so, does the outcome of environmental protection vary from case to case and why?

The analysis of _why_ environmental protection is (or is not) achieved includes both factors which could be ascribed to formal [9] requirements and procedures and factors which are not regulated or difficult to control by politicians and administrators. It is important to examine how the general forest-environmental goals are translated into specific, measurable achievements and also what strategies are used by public agencies to promote environmental protection. Is the achievement of environmental goals steerable through forest-environmental legislation? Equally important is, however, to supplement this analysis by asking the more general question of what _other_ factors, such as harvest technology or economic considerations determine the result of environmental protection.

There are a multitude of direct and indirect means by which the government may influence behaviour. As will be further discussed, both administrative and political structure and culture of the Swedish society and, especially, of the forest sector [10] are factors which are assumed to affect environmental protection. Environmentalists may lobby the public or private agencies to consider environmental aspects. Moreover, the agency officers who interpret the rules may influence environmental protection, for example through their attitudes and knowledge. Therefore, the environmental requirements of the Forestry Act are only a part of the forest-environmental policy.

Theoretical perspectives

Before my implementation analysis is presented in detail, crucial concepts and perspectives used by policy analysts will be examined. Depending on their approach, the terms "policy" and "implementation" mean different things. Implementation analysis may start from the "top" or from the "bottom", reflecting differences in basic stands towards the possibilities for a modern welfare state to identify and solve collective problems.

The "top-down" perspective examines implementation of a policy and/or of a specific programme starting with the decision made by a government agency. The "bottom-up" perspective starts with the policy problem as defined by participants in the problem-solving activities, regardless of whether they are public or private, and/or whether they are aware of being part of the policy-forming process or not.

The top-down approach

"Policy" refers to a set of political goals, defined by public institutions, which include statements of intention and predicted consequences (Pressman and Wildavsky 1973 p. xx). It may have the form of a broad statement, or of a number of specific programmes initiated to solve the collective problem (See Figure 1.1). A "programme" consists of specific governmental actions launched to secure the objectives whose attainment is problematical. Examples of government programmes include economic subsidies, regulations, rule-making and creating new authorities to achieve policy goals. A "programme" is thus not "policy" itself, but may be part of a more general policy.

Policy ⟶	Programme ⟶	Implementation of programme ⟶	Policy output ⟶	Policy outcome
(stated by a specific institution, e. g. Parliament, or the National Environmental Protection Board)	(e. g. specific legislative requirements of the Forestry Act and Nature Conservancy Act)	(performance by different actors of relevance for one or several programmes operating in the environmental field)	(e. g. degree of environmental protection achieved in accordance with forest-environmental prescriptions or Nature Conservation Plan	(e. g. degree of environmental protection totally achieved)

Figure 1.1 Relations between the terms: policy, programme, implementation, policy output and policy outcome

"Implementation" is the process of carrying out the policy or the programme(s). In other words: "Implementation may be viewed as a process of interaction between the setting of goals and actions geared to achieve them" (Ibid., p.xxi). Studying the implementation process implies investigating what happens after a policy is formulated or a programme is set up. Analysis of implementation may, however, also be made before the programme or policy is decided, with the aim of predicting how well it might be implemented. Different strategies are then evaluated in order to provide decision-makers with relevant policy alternatives (Municio 1987).

Implementation involves both political and administrative actions. The implementation process is characterized by interaction between government authorities at different levels and other actors. The study of implementation examines those factors that contribute to realization or non-realization of policy objectives.

"Policy output" refers to the policy consequences of a programme or policy. "Policy outcome" includes not only the consequences of a certain policy and its connected programmes, but also unintended - that is ultimate - results of other policies and actions. Unintended consequences of, for example, education or occupational safety policy may result in either increased or decreased environmental protection concerns. In such cases they are examples of an environmental outcome caused by programmes and policies outside the forest sector. Successful implementation of a government programme does therefore not necessarily lead to policy success. Other factors could impede or adjust actual policy outcome. Such factors may or may not be considered before policy formulation. They are, however, difficult to influence for actors in the forest sector, at least according to formal procedures.

The top-down approach to implementation analysis assumes an ideal democratic political system. In the Swedish case [11], political decisions are assumed to be rational as well as gradually specified, implemented and enforced by permanent bureaucracies. With this view, the responsibilities of the administration are fixed on different levels, and the control on the behaviour of employees performed through a set of rules within the governmental agencies. The employees simply have to follow the instructions from their supervisors, and their loyalty is ensured by a variety of review and control mechanisms (Weber 1920, 1962).

Rational policy-makers would think of objectives and goal attainment for their decided policies, and be interested in evaluating the actual implementation on the basis of intended policy objectives (Gustafsson and Richardsson 1979). Steering through programmes in a top-down manner usually assumes that society is organised hierarchically and that public administrators approximately fulfil what is prescribed by formal political decisions. Political steering, based on a top-down perspective, focuses on formal political decisions and on those implementing units specified by formal directives. Programme administration is supposed to be controlled from the top. Implementation problems emerge when administration does not adequately respond to policy directives, or when policy directives are unclear or incomplete.

Many of the early top-down researchers concluded that programmes were ineffectively implemented and formal policy goals essentially deflected during the process of implementation (Compare Pressman and Wildavsky 1973; Bardach 1977). Second generation studies analyse implementation across programmes and governmental units, comparing specific variables and conceptual frameworks (Van Meter and Van Horn 1976; Sabatier and Mazmanian 1979; 1980). Such analysis may include both conditions which refer to the initial political decisions (the "statute") and "non-statutory variables", which are the product of subsequent political and economic pressure during the implementation process.

Hierarchical control of the behaviour of lower-level officers and target groups could, with the view of Sabatier and Mazmanian (1979), be kept within acceptable bounds over time if certain conditions are met. Although other factors than the programme may be analysed when searching for explanations to the implementation process, the basic assumption of control from the top and down is not questioned.

The bottom-up approach

The bottom-up researcher views the implementation process independently of government policy and programmes. "Policy" can well exist without any programme being initiated. An investigation of the "implementation" process implies that collective actions are studied at policy delivery level (Wittrock 1985). Consequently, "policy output" is not relevant to a bottom-up researcher, since he or she does not use "programmes" or formal policy decisions as starting-points. The focus of interest is instead how different actors organise to solve a particular problem or issue. The concept of "policy outcome" and questions of what factors contribute to successes or failures are most relevant in such an analysis. Through empirical investigation, the bottom-up researcher may (or may not) discover official programmes or policies to be influential in the implementation process.

With a bottom-up perspective, the analysis begins with identifying the multitude of actors who are engaged in solving a collective problem. This collective problem is not necessarily defined by formal decisions in societal institutions, but may as well be set up by the researcher himself, based on definition by the actors at closest contact with the problem. For example, when Hjern and Hull (1982b; 1987) investigated the organisational arrangements for helping small firms grow, they defined the problem by asking the actors named as contacts by firms what they themselves perceived to be the main restriction for further development. Their study revealed a variety of private actors who perform significant intermediary roles between the firms and the actual assistance providers. The authors refer to this perspective of implementation research as "empirical constitutionalism", where all empirically relevant actors are incorporated on equal terms in the researchers' reconstruction of the policy system, whether or not they are prescribed by the formal constitution (Hjern and Hull 1982a).

The focus of analysis is, thus, on the actors who are charged with carrying out policy defined as collective action rather than with those who formulate and convey formal policy. Those who deliver services are not regarded to be at the end of the policy "chain", but are rather seen as primary actors. The local or "bottom" arena provides the context in which actors start making their discretionary judgements. In contrast to the top-down approach, this view does not presuppose a hierarchy of influence over policy to be congruent with the hierarchy of authority in the organisation (Lipsky 1980).

The bottom-up researcher puts more stress on informal than on formal rule-making. Informal rules and ethic codes among public administrators and other policy-makers are supposed to both be interesting in themselves and quite easily detected and determined when the bottom-up approach is applied.

It could be argued that the relevance of a bottom-up perspective has increased since the constitutional realities have changed in our modern mixed economy (Hydén 1984). Society reflects a multitude of organisations - both public and private - affecting policy formation and implementation (Hernes 1978). As formulated by two of the bottom-up researchers:

> Public policies today are rarely uniquely public in character; frequently they need actively incorporate private actors into the policy process in order that they may achieve their public purposes. Moreover, the constitutional presumption of neutral and efficient translation of policy goals into programme outputs must often labour under the 'deviance' of behaviour among actors at lower levels in the administrative 'hierarchy'. Who steers whom can no longer be easily assumed (Hjern and Hull 1982b, p.188).

The concept of political steering, with this view, must take into account that many key actors in implementation derive their positions and legitimacy from non-governmental sources. Control, command and aggregate measures enforced through a unified chain of command may be replaced by a strategy of negotiation, consent and persuasion (Hjern and Porter 1981). Consequently, implementation problems ought to be evaluated not from the point of view of hierarchical control from the top, but from the perspective of the actors at the point where the problem is most immediate.

Suggested synthesis and variables used in earlier studies

The main differences between a top-down and a bottom-up approach is thus whether the emphasis of the empirical work is put on analysing formal or informal implementation structures, and whether the implementation study is carried out in a basically deductive (top-down) or an inductive (bottom-up) manner. Several efforts have been made to synthesize the two perspectives and to regard them as complementary rather than mutually exclusive (Knoepfel and Weidner 1982; Elmore 1985; Wittrock 1985; Sabatier 1986; Winter 1987; Hull and Hjern 1987).

Sabatier (1986) extracts the differences between top-down and bottom-up approaches to implementation analysis in a summarizing table (See Table 1.1). The top-down approach, he suggests, is useful in cases where there is a dominant public programme in the policy area under inquiry and when a single public agency clearly dominates the policy field. The bottom-up approach, on the other hand, is more appropriate in policy areas which involve a multitude of public and private actors. The bottom-up is also better for assessing the dynamics of local variation, since it focuses on local implementation structures.

Table 1.1
Comparison between top-down and bottom-up approaches

	Top-down (Sabatier & Mazmanian)	Bottom-up (Hjern et al.)
Initial focus	(Central) Government decision, e.g., new pollution control law.	Local implementation structure (network) involved in a policy area, e.g., pollution control.
Identification of major actors in the process	From top down and from government out to private sector (although importance attached to causal theory also calls for accurate understanding of target group's incentive structure).	From bottom (government and private) up.
Evaluative criteria	Focus on extent of attainment of formal objectives (carefully analyzed). May look at other politically significant criteria and unintended consequences, but these are optional.	Much less clear. Basically anything the analyst chooses which is somehow relevant to the policy issue or problem. Certainly does not require any careful analysis of official government decision(s).
Overall focus	How does one steer system to achieve (top) policy-maker's intended policy results?	Strategic interaction among multiple actors in a policy network.

Source: Extracted from Sabatier 1986, p.33

Applying the different perspectives to the same problem leads to different results. Elmore (1979) uses the terms "forward" and "backward mapping" when he describes the methodology used by "top-down" and "bottom-up" researchers, respectively. Different factors are found to affect implementation, depending on the approach of analysis:

> The analytic solution offered by forward mapping stresses factors that tend to centralize control and that are easily manipulated by policymakers: finding formulas; formal organisational structures; authority relationships among administrative units; regulations; and administrative control (budget, planning, and evaluation requirements). The analytic solution offered by backward mapping stresses the dispersal of control and concentrates on factors that can only be indirectly influenced by policymakers: knowledge and problem-solving ability of lower-level administrators; incentive structures that operate on the subjects of policy; bargaining relationships among political actors at various levels of the implementation process; and the strategic use of funds to affect discretionary choices. The crucial difference of perspective stems from whether one chooses to rely primarily on formal devices of command and control that centralize authority or on

informal devices of delegation and discretion that disperse authority (Elmore 1979, p.605).

Hull and Hjern (1987) suggest that a bottom-up approach to implementation analysis may well be combined with addressing the central issues of co-ordination and control. Once local implementation is mapped in terms of their actor composition and strategies, the researcher is free to analyse all kinds of explanations to why particular implementation settings differ. They further conclude that:

> ...future research in the bottom-up vein will need to make a more explicit and systematic connection between the doings of the multifarious actors in the field and the actual and potential opportunities available to higher-order actors, elected politicians and senior civil servants, for structuring the behaviour of both mandated and non-mandated actors at lower levels in the policy process in order to maximise the probability that desired implementation outcomes materialise (Ibid., p.196).

In order to describe and analyse the policy implementation process, implementation variables must be defined and structured. Previous researchers have come to divergent conclusions on what variables are most important to study. The multitude of variables can be explained by differences in perspective, but also by the fact that the implementation research field is oriented towards case studies. Critical variables are often derived from the case materials (Mazmanian and Sabatier 1983). By examining over 100 implementation studies - including both top-down and bottom-up perspectives - O'Toole (1986) comes to the conclusion that there are possibilities for building some convergence in the field:

> Roughly half of the published studies identify policy characteristics (especially clarity, specificity, and/or flexibility of goals and procedures and validity of a policy's causal theory) as significant; approximately the same number claim that resources (financial and other) are crucial. Other frequently identified categories of variables include: implementing-actor or multi-actor structure, number of actors, attitudes and perceptions of implementing personnel, alignment of clientele, and timing (including the possibilities for learning among implementors) (O'Toole 1986, p.189).

Goggin (1986) makes a review of implementation research based exclusively on top-down studies. He arrives at summarizing three clusters of independent variables explaining variability in implementation performance: (1) the form and content of the policy itself; (2) the capacity of the organisation(s) responsible for making the programme work; and (3) the qualifications of the people in charge of operations, which he calls "the will, skill and attitudes of implementors". A fourth factor is also included in Goggin's schema for explaining implementation, namely (4) the environment. He does not, however, explain what is included in "the environment".

In their building of a framework for implementation analysis, Sabatier and Mazmanian (1980) discuss the influence of what they call "clearly exogenous variables", which probably correspond to what Goggin describes as "the environment". These exogenous variables are divided into: (1) socio-economic conditions and technology; (2) media attention to the problem [12]; (3) public support; and (4) attitudes and resources of constituency groups. Such groups have the capacity to affect agency policy indirectly through publishing critical mass-media reports and organising public opinion campaigns. Constituency groups can also intervene directly in the decisions made by implementing agencies, through commenting on proposed decisions and by supplementing the agency's resources (Sabatier and Mazmanian 1980).

My framework for analysis

In my study of the implementation of environmental concerns in forestry, I make use of both a top-down and a bottom-up perspective. The starting-point is the legislation and prescriptions for environmental protection issued by government authorities. The investigation concerns the results of a particular programme, namely the Forestry Act regulation on environmental protection. [13] In that sense, I work within a top-down tradition.

My analysis is, however, not restricted to an evaluation according to standards prescribed by official goals - that is policy output. In addition to these formal requirements, preservation recommendations made by flora and fauna researchers are used to establish criteria for judging the degree of environmental protection. The empirical work begins with examining policy outcome and continues by asking: What happens and why during the process when environmental considerations are attempted with clear-cuttings? Even though the policy problem is defined by formal statements, the methodology is thus strongly inspired by the bottom-up approach.

Starting with a clearfelling case study, the relevant actors are identified and questioned. There is no initial assumption that the actual policy output or outcome is controlled from the top, nor steered through the Forestry Act. Rather, I try to identify important actors by investigating, at the policy delivery point, which persons have an impact on field performance. Crucial factors contributing to the result of environmental protection are analysed through interviews with the people involved. If legislation and prescriptions made by government authorities are shown to affect the results, along with other means of steering from the top, this becomes an empirical result of the study. "Nonstatutory" factors and actors are as likely to be found influencing, or determining policy implementation as are the ones which are prescribed by formal statutes.

My analysis is based on a combination of: (1) field investigation of clearfellings; (2) interviews with implementing actors defined from the case studies, public officials working in the relevant government agencies and representatives from large forest owners; and (3) formal prescriptions and advice from the forest authorities. Methods and material used are further described in Chapter 2.

The presentation of my framework for analysis of the environmental regulation in forestry (See Figure 1.2) is made in order to organise and structure the explanatory factors and the analysis to be presented in the forthcoming chapters. I do not claim to synthesize the different approaches of implementation analysis into an overall theory. My aim is rather practical and pedagogical. I want to clarify that my theoretical assumptions are based on both the top-down and the bottom-up traditions, and to somewhat specify the variables which will be used in my forthcoming analysis.

THE SOCIETAL FRAME
Administrative and cultural features of the Swedish society: political/juridical, economic/technical
(Chapter 1)

THE FOREST SECTOR

Form and content of the environmental and forest legislation
Political and administrative culture and structure in forestry
(Chapters 3 and 7)

Independent variables

Harvest technology
Forest economic considerations

Implementing actors'
*Attitudes
*Knowledge
Environmentalists
(Chapters 4-6)

Dependent variables

Environmental protection performance/different aspects
(Chapter 4)

Figure 1.2 Framework for my implementation analysis

The independent variables listed include factors which may directly affect environmental protection policy. These factors are closely intertwined, and furthermore influenced by underlying variables - such as the forest-administrative culture and the content of the environmental and forest legislation.

The dependent variable **environmental protection performance** is supposed to be explained by the four independent variables: (1) harvest technology; (2) forest economic considerations; (3) implementing actors' attitudes to and knowledge of environmental protection issues; and (4) influence from environmentalists. The societal frame, legislation and forestry culture and structure are regarded as underlying independent variables which may indirectly affect all these independent variables.

As earlier described, forest operations have become increasingly mechanized during the seventies and eighties. **Harvest technology** is presumed to affect environmental protection both by technical conditions, such as the adaptability of machinery to different environmental situations, and by the scale of treatment. When large machinery is used, it may imply the harvest is carried out in a more stereotyped way and environmental considerations therefore difficult to achieve.

Forest economic considerations of implementing actors are likely to affect policy outcome, since there is a fundamental conflict between production and removal of as much timber as possible and protection of environmental features. Here, forest economic considerations refer to trade-off decisions between a high timber yield and environmental protection <u>after</u> the harvest technology is decided. No matter what harvest technology is selected, additional economic considerations may remain.

Implementing actors' attitudes to and knowledge of environmental protection may be of vital importance for their behaviour. It goes without saying that it is important to take into account the actors operating at the service delivery point, that is carry out the forest operations when the implementation process is analysed. Their judgements and skill, based on professional experiences, finally determine policy outcome.

Pressure from political parties, environmentalists and other organised interests of relevance to the policy problem may also affect the dependent variable. This is assumed to be the case not only during the democratic procedures when the official policy is formulated, but throughout the implementation process. For example, environmentalists may lobby the government authorities, the forest owners and the forest managers to pay attention to environmental protection. They may also succeed in attracting mass media attention to question forest-environmental outcome. In the United States, environmental interest groups and citizen opponents are found to be the principle forces to successful achievement of environmental policy goals (Andrews 1976). Therefore, influence from environmentalists is interesting to examine also in the Swedish case.

In addition to the factors already mentioned, the influence of time may affect policy implementation. During the course of time, implementing actors may learn and adapt to new programmes; environmentalist groups may become more (or less) active depending on the state of environmental protection performance; and harvest technology as well as forest economic considerations may change. Administrative structure and legislation may also be modified as a result of policy outcome.

Since I conducted the study over only a few years, eventual changes in such factors were not possible to detect. The framework for analysis shows factors which are assumed to affect environmental protection performance at a specific point of time. The framework is used for structuring analysis of variation in environmental protection between different case studies. As implied by the multitude of direct and indirect factors, policy implementation cannot be easily explained. The internal correlations between the independent variables are analysed and discussed further in Chapters 4-6.

The study is outlined in eight chapters. In Chapter 1 follows relevant background conditions of the Swedish society. Chapter 2 contains a description and discussion of methods and material used, including field investigation, interviews with forestry and government agency representatives, network interviews connected to specific clearcuttings and analysis of formal prescriptions and advice. The political history behind the form and content of the present policy affecting environmental protection interests in forestry are described in Chapter 3, which also includes an analysis of the political/administrative structure and culture of the Swedish forest sector. The role of government agencies in the policy field is discussed along with an analysis of the clarity, specificity and flexibility of legislation and procedures as specified by formal directives.

Policy outcome with clearcutting operations is examined in Chapter 4. The degree of environmental protection is analysed using a multiple regression model which takes into account eventual effects of: (1) harvest technology; (2) forest ownership (and who has fulfilled the operation); (3) regional differences; and (4) size of clearcut area. These particular factors were selected as explanatory variables in the model because they were easily obtained during the field investigation without interviews being necessary. For example, harvest technology was checked in the field and the remaining factors were determined according to the clearcut notification [14]. They were tested as expressions, or underlying variables for harvest technology, forest economic considerations, implementing actors' knowledge and attitudes and political and administrative forest culture and structure. [15]

In Chapter 5 case studies of clearcuttings are described and analysed as to what factors contributed to the result of the specific cutting. Three case studies are elaborated specifically, including high and low degree of environmental protection achieved. Implementing actors' attitudes - such as their perceptions of what environmental protection measures are important to consider - and their education in environmental protection issues are examined in Chapter 6. Influence of environmentalist groups is also discussed in this chapter as well as how implementing actors perceive the impacts of harvest technology and forest economics. Forest economic considerations and harvest technology are factors which are closely intertwined.

Chapter 7 examines the role of legislation for steering implementing actors' behaviour. The use of formal prescriptions and advice from the Forestry Agency is investigated. Different steering strategies to promote environmental protection are discussed. To illustrate such alternative strategies, comparison is made with forest-environmental legislation in the state of California, United States. Finally, Chapter 8 summarizes the results and conclusions of the entire study.

The Swedish society: A frame of reference

Before analysing the Forestry Act and the actions of the national authorities, it is necessary to clarify some characteristics of the Swedish society to an audience which is not familiar with formal responsibilities of Swedish government and its agencies. The following description outlines some basic features, which should be held in mind when the forest-environmental policy implementation is analysed.

With a population of eight million people, the majority share the same ethnical, social and cultural heritage and speak the same language. Thus, the country is relatively homogenous. [16] The State is unitary, with basically the same administrative system from one end of the country to the other. Swedish government is generally accepted and often viewed by its citizens as a "helpful father" who is willing and capable of taking care of life as a whole. There is a cultural integration of the individual and the State which implies, among other things, that there are laws for almost everything and that the State intervenes preventively in regular co-operation between politicians, bureaucrats, clients and affected interest groups (Gustafsson and Richardsson 1984).

Power to adopt new political programmes, for example in the form of new legislation, is exerted by Parliament, while preparation of government bills are made by the ministries. The ministries are relatively small, with approximately some hundred employees, led by a cabinet member. The ministries have financial and policy-making responsibilities for the several national administrative agencies with boards [17] within its domain of functional interest. Full responsibility for carrying out programmes mandated by law is lodged in these national agencies. There is an old tradition in Sweden to try to separate "political" and "administrative" decisions; and a view of administration to be carried out simply by applying the law to the facts of a specific case. The national agencies thus, with this view, are administrative and not political (Anton 1980). The traditional distribution of tasks between ministries and administrative agencies is, however, changing:

> Swedish ministries appear to be getting more involved in what were thought traditionally to be administrative matters, and administrative boards appear to wield more influence in policy-making (Elmore et al., 1986 p.230).

There are no constitutional impediments for the government [18] to exert an extensive and detailed political steering of its agencies. However, the government neither has the capacity nor the ambition to supervise and follow the agencies in detail. Swedish national agencies are relatively autonomous in relation to their ministries, which is illustrated, for example, by the fact that the General Directors of national agencies are not automatically replaced when a change of government takes place.

Instead, there are tendencies in the Swedish society contributing towards a limitation of national government control: (1) a general expansion of the public sector, especially local governments; (2) an increasing professionalism among civil servants working in the administration; (3) growing elements of laymen participation; and (4) growing influence from other societal forces than governmental, such as autonomous organisations taking over public tasks and market economy forces.

Decentralized policy-making

In several policy areas political and administrative decentralization from national level to local governments has occurred since the 1970s. Environmental protection and land-use planning are examples of policy areas where local governments are attributed considerable influence (Gustafsson

1987). Even prior to the 1970s, local governments have been largely independent from national level. According to local government legislation, the municipalities and the counties have the right to conduct "their own affairs" to the extent that specifically-regulated policy areas are not affected. Important tasks covered by this general authority concern such examples as streets and parks, cultural activities, sports, utilities etc. Policy areas which are specifically-regulated but where the municipalities are allowed to decide more in detail how to apply national policy include, for example, environment, land-use, social welfare and education. The municipality can also use its general powers to increase the extent and/or quality of goods and services prescribed by national and/or regional authorities.

By tradition, local government has played an important role in the Swedish society. The 284 municipalities and 23 counties [19] are elected bodies with power to tax their inhabitants and provide services as well as to interpret national policy recommendations in their field and make them applicable to local circumstances. The limit of what is regarded to be "their own affairs" is judged by the citizens and their representatives. Only those living in the municipality or county in question have the right to appeal against local government decisions. Therefore, the range of activity within the local governments is largely determined by their own citizens, and not only by nationally defined legislation. Generally, municipal tasks have increased during the last two decades (Kolam 1987).

The expansion of the public sector has led to: (1) removal of some administrative matters from the Cabinet to subordinate authorities and creation of new authorities; (2) a transition from steering through detailed legislation to softer steering often through frame-laws; and (3) development towards greater economic autonomy for administration through frame-budgeting (Royal Commission Report 1983:39). These tendencies are largely uncontroversial from the point of view of party politics in Sweden. Since the government steering of environmental protection considerations with forest operations is based on such a frame-law, this phenomenon has to be further elaborated.

Frame-laws

In Sweden political goals are often manifested in terms of frame-laws. During the last twenty years, an increase in the number of frame-laws has occurred. Frame-laws are used for defining general political goals; yet they are incomplete in that they are assumed to be supplemented by additional regulation. The use of frame-laws instead of traditional, detailed legislation, has been motivated by the constant, rapid changes of modern societies like Sweden. The rules must therefore continuously adopt to societal changes. Complexity in societal organisation is another motive for the use of frame-laws. It has generally been regarded impossible to regulate in detail how different political goals should be balanced. Specification of regulation has therefore gradually been transferred from Parliament to the bureaucracies of the national agencies and to local governments (Hydén 1984).

Figure 1.3 describes the formal responsibilities of different levels of government with a frame-law system.

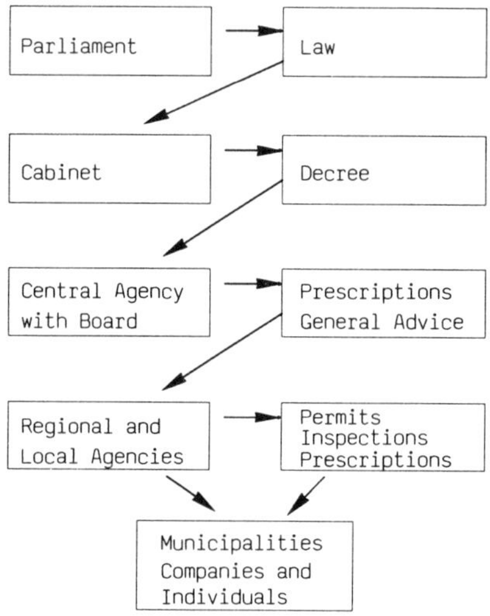

Note: Translated from the National Accounting and Audit Bureau. RRV Dnr 1979:1199 (from Hydén 1984, p.9)

Figure 1.3 Institutional responsibilities of rule-making with a frame-law system

The amendment of law is made in Parliament. Frame-laws may be specified through prescriptions only if this is explicitly required in the law. Hence, Parliament may delegate the right to formulate additional rules to the Cabinet as well as to national agencies. Regulation in detail may thus be made by civil servants. Political motivations in a Bill preceding a decision in Parliament also serve as a guide when agencies produce or implement prescriptions based on the frame-law.

The transition from detailed legislation to frame-laws involves the administrative agencies in political decisions. Vague expressions and abstract rules in the frame-laws necessitate interpretation. The implications of a frame-law have to be specified and conflicting interests balanced. Efforts to achieve this are made through participation of different interests in the rule-making process. Generally the agency forms a committee consisting of agency officials and other representatives whom it has selected to work out the text of a regulation.

Comparison between Sweden and the United States

Conflicting interests are solved differently in Sweden compared to, for example, the United States. In the United States the outcome of conflicts is settled in court. Like in Sweden, potential conflicts are assumed to be mitigated through policy formulation. The main difference is the amount of time and effort spent during policy formulation. In Sweden, constitutional processes require extensive formal consultation among all parties affected by a policy before it is enacted. Potential conflicts are therefore, presumably, stabilized before the policy comes into practice. The two main structures of consultation are Royal Commissions, expert advisory boards on key policy questions, and the "remiss" process - an established routine by which the ministries solicit detailed comments from administrative agencies and interest groups.

Proposals submitted for legislative decision after remiss include a consideration of how the proposed Bill has taken account of criticism. After policy formulation, there are few mechanisms in Sweden compared to the United States to correct eventual remaining conflicts. In addition, Swedish political parties tend to tone down remaining conflicts after policy is formulated rather than, as is common in the United States, emphasizing them for electoral purposes (Elmore et al., 1986).

The characteristic implements of Swedish policy - structured consultation and planning - are consistent with a governmental structure that assumes a basic compatibility of interests among levels of government. In the United States, as a contrast, the characteristic implements of policy - delegation, distribution, rights, and adversarial procedures - are consistent with a governmental structure that assumes a basic conflict among levels of government (Ibid.).

The consensus-making role of Swedish bureaucracies is well-known; the politics of compromise is firmly rooted in Swedish tradition (Rustow 1955). In a comparative study of occupational safety and health Kelman (1981) analyses major differences between Sweden and the United States. In Sweden the same values and institutions are used during rule-making as are applied for compliance with the rules. In these negotiations people often come to appreciate the perspectives and values of those with whom they are in conflict (Kelman 1981). A comparative study by Lundqvist (1980) of clean air policies in Sweden and the United States confirms the differences in policy strategies. The policy-maker in the United States seeks radical policy alternatives as an answer to public opinion pressures, and the balance of power is determined afterwards in the courts. In Sweden, the policy-maker is inclined to have both the governmental institutions and the regulated interests move at the same speed and in the same direction. The speed at which policy is developed is also adjusted to the resources available (Lundqvist 1980).

Consequently, the steering strategy used in Sweden is characterized by "soft" methods. There is a general reluctance to use punishment and force. As Lundberg (1982) states in his study of the Swedish Work Environment Act implementation, the government agency in charge stresses information and knowledge, and enforces the act through voluntary compliance. The inspectors try to convince the employer of the benefits of safety and health by argument and advice. Sanctions are rarely used. Instead, the inspectors seek to establish a friendly, personal relationship with the people at the site.

The ideal civil servant holds "soft" qualities rather than the formalistic insensivity often attributed to the bureaucratic role:

> He must be aware of the needs of his subordinates, be willing to participate rather than dictate, and he must be equally sensitive to the problems of his peers, with whom he negotiates. He must, above all, be able to get along with people (Anton 1980 p.75).

At the same time, the ideal-type administrator is seen as an "activist" rather than a pure executor of orders from above:

> He must have a sense of purpose, awareness of the opportunities available in the environment, and the ability to sell both himself and his program plans (Ibid., p.75).

However, the ideal model of consensus, compromise and long-term orientation is to some extent changing in reality; the behaviour of Swedish agencies is increasingly characterized by uncertainty, complexity and incremental decision-making through negotiations. There is diffusion of power and symbolic policy ingredients have developed (Gustafsson 1987).

NOTES

[1] These are approximate figures developed from Alén (1977), Ahlén et al., (1979), Ingelög (1981), Ingelög et al., (1984), and Ehnström and Waldén (1986). The figures refer to "threatened" species, which include "immediately endangered", "vulnerable", "rare" and "requiring special consideration", and have been

earlier presented in Pettersson (1986).

[2] Mechanization in Swedish forestry has increased from 16 per cent in 1972 to 63 per cent in 1982, referring to the share of time spent by chain saw workers replaced by more mechanized methods (Statistical Yearbook of Forestry 1987).

[3] However, use of herbicides in forestry has drastically decreased during the 1980s due to new legislation.

[4] National parks, and nature reserves (including Swedish Forest Service reserves, so called Domänreservat) comprise 4.5 per cent of Sweden's total land area. Protected forests within these reserves amount to approximately 0.5 per cent of the total forest land base. In some of these protected forest areas cutting is, however, allowed to a certain extent (Minutes from the National Environmental Protection Board, 1987).

[5] "Economic forester" refers to a land manager whose principal interest in forestry is the highest immediate financial return from timber production (here called "economic forestry"). An opposing principle goal would be "multiple use forestry", i.e. when the total benefits of wildlife, recreation, berries and mushrooms, water quality, flora and fauna protection are considered as well as timber production. Multiple use forestry can obviously in a broader perspective be regarded as economic forestry - the difference is that it includes social economics. "Social economic forester" refers to a land manager with multiple use goals.

[6] "Study circles" are a typical Swedish social phenomenon, where a small group of people gather to study a specific topic. One person acts as a group leader, often because of his/her knowledge and experience in the subject of study. The members of the study circle are provided with literature, which they study in between meetings. During the meetings they discuss their findings, usually by guidance of prepared questions. Study circles are often carried out in the evenings, but are also promoted by the labour union and arranged during working hours. During working hours the deduction in salary is compensated by a grant from the Labour Union Study Organisation (Arbetarnas Bildningsförbund).

[7] A calendar with environmental protection information was sent to all members in the Forest Workers' Labour Union in 1984 along with a request to care for the environment as outlined by the Forestry Act.

[8] "Clearcutting" and "clearcut" refers to a final felling of a forest area. Seed trees may, however, be

left for regeneration. Approximately 25 per cent of Swedish forests are naturally regenerated through seed trees.

[9] "Formal" decisions, or rules, refer to that they should - as they are defined here - be regarded as authoritative and official by the general public. The opposite is informal decisions or rules which may be regarded valid for a group of people with the same educational background or interest, and which are not official - in a public debate they are presented as personal views (Gustafsson 1987, pp.16-7).

[10] "Forest sector" is here used as a term for the private and public forest organisations and actors involved with forest public agencies who supervise forest activities according to forest and environmental legislation.

[11] A further description of the typical Swedish political system is found in Rustow 1955, Anton 1980, Kelman 1981 and Elmore et al., 1986. A short review also follows in the last section of this chapter.

[12] In a later version of the framework for implementation analysis, media attention to the problem is included in the variable "public support" (Sabatier and Mazmanian 1983).

[13] Clearcuts are chosen to represent forest activities regulated by the Forestry Act both because of practical reasons and because the environmental impact is considered to be great. The practical reasons are elaborated further in Chapter 2, where I describe and discuss methodology and material.

[14] Further information on clearcut notifications follows in Chapter 3.

[15] Relations between the factors used in the multiple regression model and the variables in the framework for analysis are further described in Chapter 4.

[16] During the last two decades, Sweden has become more heterogeneous due to, for example, immigration from other cultures. Still, Sweden is homogeneous compared to most other countries.

[17] Several of the quoted authors refer to the Swedish "ämbetsverk" as "administrative boards" (Compare Anton 1980, Elmore et al., 1986). I prefer to call them "national agencies" since they are large organisations with many hundres (or thousands) of employees, however, led by a laymen board. This board decides in principal matters, but the everyday administration is carried out by civil servants. In the text, "administrative boards" are used with the same meaning as "agencies".

[18] The "government" is here generally used as a term for both the Parliament and the Cabinet.

[19] There are actually 24 counties in Sweden, but one of them (Gotland) constitutes a municipality which also covers county tasks.

2 Methodology and material

The study combines field investigation of clearcuttings, interviews with implementing actors and evaluation of Forestry Agency documents on environmental protection issues. Interviews were carried out both through network analysis starting from case studies of clearcuttings, investigated in the field, and with different forest and environmental organisations. As mentioned in the introduction, the methodology used contains both a top-down perspective - an analysis of formal institutions and their methods of steering - and a bottom-up perspective - beginning with actual policy outcome and behaviour of field level implementors. Hence, the methodology used does not assume that the policy initiated by legislators is the only - or even the major - influence on the behaviour of people during the policy process (Elmore 1979).

The use of multiple sources of evidence - a major feature of this study - is common in implementation research. When Yin (1982) examines the methodology used in eleven exemplary studies, he finds multiple sources are used in all of them. Two different purposes are achieved with the use of multiple sources. Yin explains this further:

> First, multiple sources were used within a single case study to demonstrate a certain degree of consistency and corroboration about implementation events and interpretation (internal validity). If information from an unstructured discussion is consistent with information from existing documents or field observations, for instance, the researcher's interpretation is more credible. This is not to say that implementation events have an objective truth; specific 'facts of matter' may

differ, depending of the perspective of the particular actor. However, the multiple sources can and should be used to support the existence of these different perspectives, and even to explain them...

The second purpose for using multiple sources of evidence was to extend the scope of study to cover more sites (external validity). As mentioned, multiple-site studies can be based on more intensive investigation of a few sites, combined with formal interviewing or other methods to cover (more superficially) a larger number of sites. The intensively investigated sites usually provide that basic information for interpreting and explaining the implementation process; the superficially investigated sites allow the investigator to expand the generalizability of the findings (Yin 1982 p.50).

Methodology and material used in the analysis is presented as follows: (1) the field investigation; (2) the preparatory interviews with different government agency and forest representatives; (3) the network interviews based on cases of clearcuttings from the field investigation; and (4) the evaluation of formal prescriptions and advice from the Forestry Agency.

Field investigation of clearcuttings

The Swedish Forestry Act's environmental protection requirements are applicable to most forest management operations, such as regeneration, spacing and thinning, clearcutting and road-building. All these measures affect nature protection. If, for example, nesting trees are removed in a thinning operation, a valuable wildlife habitat will be lost - and a potential environmental protection element will be eliminated before clearfelling takes place. Clearing, spacing, thinning and road-building are carried out without application or permission from the government agencies, which makes it difficult as a researcher to select forest areas for evaluating compliance to environmental guidelines. Since clearcuts must be reported at least one month in advance to the County Forestry Agency, selection of areas can easily be made.

Similar investigations have been hard to trace in other countries. Possibly, evaluation is made within the forest organisations or within government agencies and not officially published. One study in the United States, carried out in the Adirondack region of New York State, resembles the field evaluation of the Swedish clearcuttings (Monteith 1978). Researchers examined the extent of voluntary compliance to New York's Timber Harvesting Guidelines within thirty clearcut areas. The environmental and aesthetic aspects of major travel corridors, access roads, skid trails, landings and stream protection zones were evaluated according to a four-graded scale: "complete", "substantial", "moderate" and "incidental" compliance. The overall logging performance was found to be in compliance with the guidelines. A similar methodology has been used in this Swedish study.

Location and design of study

Proposed clearfellings were selected from the applications at the County Forestry Agency, with neither the Forestry Agency officers nor the forest owners knowing which areas were evaluated. A stratified random sample of clearcut notifications represented different forest owner categories in different counties. An important advantage with this methodology is that environmental conditions could be investigated on the forest area before cutting commenced. Already clearcut areas would have been very difficult to evaluate if pre-harvest conditions are not documented. However, investigation both before and after clearcutting is very time-consuming and the sample thus heavily restricted by research funds.

In the beginning of the study, the selection of counties included two northern counties (Västerbotten and Västernorrland) and one southern (Östergötland) (See Appendix 1). In addition, two more were investigated - one in the very north (Norrbotten), and one in the middle of Sweden (Värmland). The two counties added were investigated by university students, whereas I carried out the main part of the field investigation in the initially selected three counties, with the help of an assistant. Since the same method was used, their data were incorporated into the analysis. The choice of counties was made to represent different forest conditions, to reflect social aspects of environmental protection, compare administrative routines and (for practical reasons) to increase efficiency.

The field investigation of clearcuttings was carried out during four successive harvest seasons. During 1981/82 clearcuts in the counties of Västerbotten and Västernorrland were inspected. Investigation continued in Värmland in 1982/83, Västerbotten, Västernorrland, Värmland and Östergötland in 1983/84 and Norrbotten in 1984/85. It is not possible to evaluate whether any change in environmental protection took place during the period, since the study was made in different counties at different times. In the counties of Västerbotten and Västernorrland the evaluation was made by two different evaluators during the two seasons, making the influence of time impossible to separate from divergence in evaluation by different people. Comparison between different evaluators revealed no significant tendencies in judgements (Eckerberg 1986).

The sample of clearcuts was made in proportion to the

ownership share of "private" (individual, non-industrial), "private companies" (industrial), "Forest Service" (state-owned) and "other public forests" (the Church, municipalities and common forests) in the different counties. Therefore, the number of investigated areas among different forest owner categories varied between counties. A total of 216 areas were examined before clearcutting (See Table 2.1).

Table 2.1
Clearcuts investigated before (in parenthesis) and after felling, by county and forest owner category

Forest owner category	Norr-botten	Väster-botten	Väster-norrland	Värmland	Öster-götland	Total
Private	(11) 8	(26) 20	(17) 13	(17) 13	(37) 32	(108) 86
Private companies	(5) 3	(10) 8	(24) 15	(13) 11	(10) 9	(62) 46
Forest service	(14) 2	(13) 10	(6) 4	- -	(6) 6	(39) 22
Other public forest	- -	- -	- -	(1) -	(6) 4	(7) 4
TOTAL	(30) 13	(49) 38	(47) 32	(31) 24	(59) 51	(216) 158

Twenty-seven per cent of the areas investigated before clearfelling were not harvested during the time proposed according to the applications. Most of these areas were found within large-scale forestry, especially the Forest Service (See Table 2.1). According to the interviews, forest owners might change their mind or deliberately apply for too many clearcuttings in order to provide room for planning changes. The extent of environmental features present before harvesting did not differ between harvested and non-harvested areas (neither within counties nor within forest owner categories). Therefore, the sample may be considered representative for "average Swedish forestry" and non-biased (Eckerberg 1986).

Field procedures

Field inspection was designed so that different evaluators were able to make the same judgements. A check list of environmental protection measures was developed from the NBF prescriptions in the Forestry Act § 21 and from preservation recommendations made by researchers. [1] The evaluation of environmental protection achieved includes only relatively small measures, which are to be considered on all forest land, and does not refer to the total concerns for protecting endangered and threatened species. In accordance with the Forestry Act regulation, criteria used in the investigation are compatible with economic forestry and do not imply retention of large forest areas. A total of forty-four categories of environmental protection measures were listed. [2]

Before clearcutting, the forest area was cruised systematically, both by a grid system of 50-100 metres spacing and by searching parts of the area expected to be environmentally important, such as lakes, watercourses, bogs and hills. The different environmental protection measures to be considered during a clearcut operation were recorded and described prior to the harvest.

In addition, photographs were taken when possible. The location and the direction of photographs were described on the map, and marked in the terrain with a plastic stick, in order to facilitate a follow-up after clearcutting. The photographs were used primarily as reference material. They were later used as a basis for discussion during the interviews. In addition, they were useful to document examples of environmental protection.

Some variables were difficult to detect, or they rarely appeared, for example birds of prey's nests, capercaille courtship areas, habitats of rare plants and cultural resources. In such situations, the "check-list method" was not appropriate. It would have been necessary to sample the areas investigated according to the frequency of specific environmental protection features. The method used in this investigation sampled areas representative for the average Swedish forest land, rather than forest areas which require special environmental considerations, since the areas were selected from different forest owners regardless of environmental values according to Nature Conservation or Forest-Environmental Management Plans. [3]

After the harvest, an evaluation was made for every pre-harvest measurement. A three-grade scale was established: "full consideration" implied cases where all, or substantial, parts of the environmental protection measure were retained; "partial consideration" inferred at least some features were retained; when nothing, or very little, of the environmental protection measure in question was cared for, it was judged "no consideration".

Discussion of method

There is a problem when the requirements of the Forestry Act regulations are to be translated into specific, measurable achievements. Although many examples of environmental protection measures are mentioned in the regulation, the expressions as to what extent the protection is required leave room for different interpretations. Even if some environmental protection measures are only considered to "partial", or even "no" extent, the total result of environmental protection of the clearcutting might be adequate.

The validity of the method was tested by an alternative definition of "degree of environmental protection". Environmental features were quantified by an independent measurement and then compared to the initial judgements. The alternative method gave a "grade" of environmental protection based on a total judgement of the clearfelling operation. With a scale of 1-4, equivalent to "failure", "failure with some hesitation", "passed with some hesitation" and "passed", the compliance of the Forestry Act regulation was evaluated (See Table 2.2).

Table 2.2
Comparison of 82 clearcuts with "grades" of environmental protection according to two different methods of judgement

Grade	A total judgement (testing method)	Percentage of environmental consideration (the method used in the investigation)	
	Share with this grade	Share with this percentage of consideration	Summary of consideration
1 "Failure"	15%	15%	0 - 25% consideration
2 "Failure with some hesitation"	22%	27%	26 - 50% consideration
3 "Passed with some hesitation"	46%	39%	51 - 75% consideration
4 "Passed"	17%	19%	76 - 100% consideration
	Average "grade" = 2.7	Average "consideration" = 56%	

The evaluation method used in the study calculates a quotient of environmental consideration for each clearcut area, which may be expressed as a percentage. This quotient is calculated by a summary of "full consideration" = 1, "partial consideration" = 0.5 and "no consideration" = 0 for every environmental feature present, divided by "possible environmental protection", that is the total number of environmental features of each area:

formula (1)
$$\sum_{i=1}^{n} \frac{B_i - A_i}{n}$$

n= number of environmental protection measures relevant at the clearcutting

B= points before clearfelling

A= points after clearfelling

The two methods of judgement are thus compared for 82 of the clearcuts investigated. [4] When the same person is judging, the correlation between the two methods amounts to r=0.736. The limit for "passed with some hesitation" corresponds to approximately 50 per cent consideration of the summary calculation for each clearfelling. "Passed" is equivalent to 75-100 per cent consideration. This test of validity is however not exhaustive. It is, naturally, impossible to free oneself totally from the first judgement method when applying the "test method". Therefore, the two different methods of judgement are intercorrelated.

Reliability of data was tested by letting representatives of the County Forestry Agencies, the National Board of Forestry, the Environmental Protection Sections of the County Administrations and the National Environmental Protection Board carry out the same judgements of degree of environmental protection. Five clearfelled areas in the county of Västernorrland and eight areas in Östergötland were re-evaluated. The areas represent different forest owner categories and a variety of environmental protection shown, ranging from "no" to "full" consideration. The new judges were informed on site of what the forest looked like before cutting, using photographs and descriptive notes from the earlier investigation. After a tour of the area the representatives were asked to pass in their judgements, not knowing the original evaluations. It is difficult to totally avoid influence from the original judgement, since all aspects of environmental protection present before cutting cannot be objectively described. However, if the new evaluator found some environmental protection measures were not necessary, he could delete them from the record.

The difference between new and original evaluation gives a picture of the variation between different judges, as well as of the reliability of the method. For every measurement made, deviations were quantified from zero (= no deviation) to two points (= maximum deviation), and summarized for each clearcut respectively (See Table 2.3). If the new judge was more "prone to protection", a negative figure was obtained, while a positive deviance meant he was more directed towards "economic forestry".

Five persons made 101 judgements each in Östergötland and four persons made 55 judgements each in Västernorrland. The sum of deviations of all judgements amounted to +54, or with absolute figures 197. The largest possible deviation could have been -823, if the evaluator was an extreme "environmentalist", and +689 if he was an extreme "economic forester". (The deviation figures are not equal as to positive and negative deviation respectively, due to the original judgement.) The largest possible absolute deviation of the areas re-evaluated amounted to 1,250. Hence, the new judges were around 8 per cent more directed towards economic forestry than the original. The absolute deviation, that is in both directions, was approximately 16 per cent of the total possible compared to the original evaluation.

It is important to analyse the protection features in which these deviations occurred. The largest difference was present with clearcuts # 1, 7 and 11. Actually, the deviation was caused by very few judgements. For example, at clearcut # 1 a forest path was covered with logging debris when the area was first inspected, but at the time of re-evaluation the debris was cleared away. In other words, there had been a change at the clearcut between the two evaluations. With clearcut # 7 there was a fen with alder trees (Alnus glutinosa), where the new judges were more tolerant of it being harvested. At site # 11 the difference in judgement concerned cleaning up debris along the shore and along a forest path. Originally, these environmental protection measures were judged as "partial consideration" shown, but the new judges called it "full consideration". In some cases the deviations could, accordingly, be quite large.

However, disregarding the judgements mentioned above, the total deviation was only +14, or 2 per cent more economic forestry oriented than the original judgements. In absolute figures the deviation diminishes to 158, corresponding to 13 per cent of the possible absolute deviation. There were no significant tendencies as to what categories or aspects of environmental measures were differently judged. The control made by forest and environmental agency representatives thus shows approximate agreement between different judges as to what is required by the Forestry Act.

Table 2.3
Difference between original and new judgement
among objects and agencies

County	Clearcut #	Forestry Agency officers (3 persons)	Environmental agency (2 persons)	Total	Total possible positive difference	Total possible negative difference
Östergötland	1	+11	+6	+17	+80	-30
	2	0	-6	-6	+50	-80
	3	0	-7	-7	+35	-125
	4	+3	0	+3	+45	-145
	5	+3	0	+3	+45	-125
	6	+2	0	+2	+60	-40
	7	+21	+14	+35	+90	-50
	8	+1	0	+1	+64	-16
		(2 persons)	(2 persons)			
Västernorrland	9	+3	0	+3	+16	-32
	10	+1	-2	-1	+16	-56
	11	+8	+4	+12	+56	-72
	12	-2	-3	-5	+56	-40
	13	-2	-1	-3	+76	-12
TOTAL		+49	+5	+54	+689	-823

The investigation evaluated the environmental protection achieved immediately after the harvest season. Some features could, however, change over time. On one hand, soil scarification [5] might affect the environmental values negatively, leading to additional deep tracks or compacting residual vegetation at the clearcut, and post-harvest draining of wet areas might destroy living conditions for different organisms. On the other hand, the results could have improved the area if deep tracks were restored and logging debris taken away from forest paths and watercourses.

The explanatory model

The analytical model of environmental protection with clearcuttings is limited to factors possible to determine at the time of field inspection. The model is described by the function:

formula (2) Y = f(harvest operation system; who has fulfilled the operation; county; size of clearcut area)

where Y is the dependent variable "degree of environmental protection", which is to be explained.

In order to estimate the influence of the independent variables the dependent variable must express independent observations. Since different environmental features of one clearcut area are inter-related, they have to be weighed together in the explanatory model. The dependent variable used in the model is the "total degree of environmental protection". The dependent variable is also split into different aspects such as "technically and economically motivated [6]", "aesthetic" and "pure floristic and faunistic" environmental protection as well as "cleaning of debris and avoiding deep tracks".

The dependent variable consists of a quotient ranging from zero to one, according to formula (1). The independent variables are treated as dichotomies, so-called "dummy-variables". For example, the variable "harvest operation system" was valued "one" for mechanized harvesting and "zero" for all other harvest operation systems. The only independent variable expressing an interval scale was the size of clearcut area, which varied from 1-152 hectares in the material.

One criteria for using linear regression (OLS [7]) when analysing the explanatory power of the independent variables is that the dependent variable must consist of an interval scale. However, the "degree of environmental protection" used here does not correspond to an interval scale. It often assumes the value "zero" or "one" and it can neither assume negative values nor exceed the value "one". It is therefore necessary to use a method applicable to when the scale is truncated and, consequently, the residuals not normally distributed. By using a "probit model" the values of the dependent variable are transformed into the normal probability function and thus the residuals are also adapted. Thereafter, it is possible to estimate the impact of independent variables by maximum likelihood methods (Pindyck and Rubinfield 1976; Maddala 1983).

The values of the 158 clearcuts investigated are analysed with multiple regression. It is important to be sure of that the different independent variables are not highly inter-correlated, since such a multicollinearity can affect the standard deviation of the estimated regression coefficients. Correlations were tested and found to be moderate according to normal standards (Eckerberg 1986). [8] Some of the variables with the highest correlations were tested further in the analysis in order to check the results. For example the private forest companies used more mechanized equipment when cutting on their own forest land and the private, individual forest owners cut more manually compared to the average - a situation which affects the interpretation of multiple regression results. The analysis of the field investigation of clearcuttings is continued in Chapter 4.

Preparation

In order to become familiar with the object of research, it is necessary not only to study available documents and literature, but also to discuss with people who know the subject. During such interviews the questions for further research are brought up, and additional questions formulated. After only a couple of months with the research project, I attended four conferences (during 1981) arranged by the National Environmental Protection Board and the National Board of Forestry. The aim with these meetings was to clarify responsibilities and co-operation between the government agencies with forest environmental protection.

A preparatory conference consisted of nineteen representatatives from national and county agencies; the three following were regionally arranged, each with around thirty administrators participating from the Environmental Protection Sections at the County Administrations and the County Forestry Agencies. In addition, the cultural protection interests were represented through the National Board of Antiquities or the Regional Inspector of Antiquities at the County Administration and the municipal interests through the Swedish Association of Local Authorities. The conferences included field trips to clearcuttings where environmental conflicts were discussed. The conferences gave an overview of similarities and differences in administrative routines between the different counties. Questions of conflict were also brought up, as well as different strategies to achieve better environmental protection results. [9]

In spring 1981 a total of nineteen persons were personally interviewed. Respondents were selected among EPS and CFA officers in three counties [10] and among large forest-owners such as the Forest Service and two forest companies. The majority of persons questioned worked at the county or regional level; two District Forest officers, one forester and one timber officer represented the personnel at the closest point of environmental service delivery in the field. Prepared questions were put to the respondents; much of the answers covered not only these questions but also other aspects of implementing environmental protection in forestry, which were brought up during the interviews.

I visited the headquarters of both the National Board of Forestry and the National Environmental Protection Board. The administrators responsible for environmental protection in forestry were interviewed. Three of them had participated in the formulation of the Forestry Act regulations on environmental protection, why questions especially referred to the ex ante assumptions on practical application of the law.

Environmental interest groups were also heard. For example, I attended a seminar in spring 1981 arranged by the Ornithological Association of Västerbotten and Norrbotten counties, where the Forestry Act environmental regulation was discussed. In addition to the twenty-five ornithologists, a local representative from the Forest Service and two District Forest officers were invited to the seminar. Participants reported on perceived conflicts with economic forestry and on different ways of improving environmental protection measures through collaboration with forest owners and government agencies.

During the first three years of the research project, an advisory committee was tied to the project, consisting of one representative each from the NBF, the NEPB, the County Forestry Agencies and the Environmental Protection Sections of Västerbotten and Västernorrland counties. This advisory committee was formed on the initiative of the NEPB, who sponsored much of the project. I gathered the group once a year to give comments on methodological questions and to discuss preliminary results.

I published results of the field investigation of clear-cuttings already in 1984 (Eckerberg 1984; 1985). Since the issue of environmental protection in forestry is very much on the political agenda, these results were spread to a large audience and discussed in mass media. I received frequent letters and calls, and was often engaged in seminars and courses on forest environmental protection, arranged by forest owners and managers as well as by environmental groups. Through these contacts I received a lot of information which I was able to use in my analysis.

Network interviews

By interviewing people in a network analysis it is possible to identify those actors who are important in determining the outcome of policy at field level. The idea of an "implementation structure" as a unit of analysis is developed by a group of researchers at the International Institute of Management in Berlin. These implementation structures are defined from the linkages at the bottom level and do not necessarily coincide with formal administrative and organisational hierarchies (Hjern and Porter 1981).

Starting from a specific clearcutting, the people involved were identified. One interview led to the other by simply asking who was an important contributor to the cutting. There are several advantages of having a specific case to refer to during the interview. A researcher is certain to be able to reach the <u>real</u> operatives in the implementation process and it is possible to specifically question what happened and why. With this method, it is difficult for the respondent to answer in a "wishful-thinking way" (i.e. telling what should have been done as opposed to what actually was done), which is judged to considerably increase both validity and reliability.

The choice of personal interviews instead of a mailed questionnaire or telephone interviews, is based on the nature of the research object; it takes time and a sensitive ear to put questions about the implementation process, if the answers are to be reliable and not only expressed to meet presumed expectations from the researcher or from superiors. The use of structured open interviews leads to un-biased responses, and they encourage explicit answers from the respondents. However, the researcher's influence cannot be totally avoided, even if it can be minimized through certain efforts (Murphy 1980). In addition, there is ample time during the interview to elaborate on the specific issues for each case, thereby generating answers that might not have been predicted. There is also a possibility to follow-up any statement which does not correspond with other equivalent responses from the same clearcutting case.

A sample of twenty-five clearcuttings, which had been investigated in the field, were examined by network interviews. The twenty-five cases were selected from four different counties (Västerbotten, Västernorrland, Värmland and Östergötland), representing different forest owner categories and performed by different harvest operation systems. In addition, the clearcuts were selected to represent clearcuts with high degree of environmental protection achieved, as well as clearcuts where little consideration was shown. The selection of case studies was made with the aim of identifying important determinants to the result of environmental protection, and to illustrate what happened and why at a specific clearcutting. The case studies do not represent the average Swedish clearcuts, since they are not selected at random. Still, the case studies may be used to picture the implementation process, to formulate hypotheses, and to analyse causal relationships.

Starting from a name and address on the clearcut notification sent to the County Forestry Agency, all the important participants during the harvest operation and the forest management planning were identified: the forest owner, the forester, the timber officer, the forest workers and the District Forest officer. Except for a few cases all these actors were interviewed. (Sometimes it was impossible to arrange a meeting of practical reasons. [11]) None refused to participate in the study. However, all forest owners were not interviewed, only the private individuals and representatives of common forests. Two reasons made me decide not to interview the higher officials or the real owners of the forest companies, the state forest and the municipality-owned forests. First, economic resources restricted the number of interviews that were possible to accomplish. Second, the implementing actors "more close to the field" were more important to interview, with my bottom-up approach of analysis.

A total of 115 forest professionals [12] were interviewed. The number of interviews for each clearcutting case ranged from two to eight, with an average of 4.6 persons per case study. The interviews were conducted during a period of two years. A pilot study, with twenty-two interviews, was made in winter 1983/84. The remaining interviews were carried out in autumn 1984 and in the beginning of 1985. The employer and the classification of each respondent is described below (See Tables 2.4 and 2.5). [13] Almost one-fourth of the sample were Forestry Agency officers, whether or not they actually directly influenced the harvesting. [14] The category "others" consisted of two persons of whom one worked with forest workers' safety, but had a job on the side as forest consultant to private forest owners, and the other was a forest supervisor (in the administrative hierarchy above foresters and timber officers) at a private forest company. Since the category "others" only represented two persons, they are excluded in the tables in Chapter 6, except when all categories are summarized.

Table 2.4
Respondents' employment

Employer	Number	Percent of sample
Swedish Forest Service	15	13
Private forest companies	40	35
Forest Owners' Association	16	14
Forestry Agency	26	23
Private enterprise	10	9
Forestry society	6	5
Other	2	2
TOTAL	115	100

Table 2.5
Respondents' category of work

Category of work	Number		Percent of sample	
Forest workers	45		39	
- manual (chain-saw)		18		16
- felling machines		4		4
- harvester		18		16
- forwarder		5		4
Timber officers	23		20	
- foremen		19		17
- planners		4		3
Foresters	11		10	
Forest owners	10		9	
- work in the forest		8		7
- do not work in the forest		2		2
Forestry Agency officers	24		21	
- district advisors		21		18
- other tasks		3		3
Others	2		2	
TOTAL	115		100	

The length of the interview ranged from thirty minutes to two hours, depending on the respondent. Most of the interviews were carried out with one person at a time; sometimes group interviews had to be made for practical reasons. [15] In the latter case the interviews were at the place of work. Many of the interviews took place in the respondent's home. I did not, however, perceive any differences in attitudes towards the interview situation depending on the location of the interview. The only problem occurred during the group interviews where usually one or two of the respondents answered most questions. It was impossible to know if the others would have answered in the same way if they had not heard the previous answer. I registered one answer for every person whether or not everyone actually responded during a group interview. Silence was viewed upon as an agreement. [16] When there were discussions where different participants claimed different opinions, the answers were registered for each person respectively.

The interview schedule consisted of the following: first, a brief description of the aims and methods of the research project (without presenting the results), then questions about the specific clearcutting were asked, followed by general questions. Finally I asked questions about attitudes, education and memberships. By starting with the specific clearcutting already investigated, it was easy to turn over to general perceptions of implementation aspects. The interview schedule is shown in Appendix 2.

Occasionally, not all the prepared questions were discussed during the interview. The interview was more of a structural discussion than a "question and answer" dialogue. When respondents asked for my opinion I refrained to make any statements until the interview concluded. All interviews were tape-recorded [17]; later the recordings were transcribed. The answers were coded and analysed. Some of the questions were open-ended and the respondents did not always answer questions exactly according to the interview schedule. For example, a respondent may have discussed economic considerations when he explained the result of the specific clearcut, but when he was later asked about the influence of economics he could not remember any case where economic considerations determined environmental protection measures. Therefore, each respondent's answers were analysed as to internal inconsistencies, and interpreted as a whole into groups of answers to compare, for example, different categories of workers. [18]

Forestry agency prescriptions and advice

The only way the Forestry Agency can enforce environmental protection requirements is through issuing a prescription which specifies what measures must be taken at the specific clearcutting. Influence may also be exercised through advice given by the Forestry Agency officers. Prescriptions are always formalized in documents; advice, however, may be either oral or written. From 1980 (when the new forest environmental legislation was amended) to the end of 1986 there were twelve such "Prescriptions" and 2,424 "Instructions and Advice" issued nationwide (on standardized forms). With the aim of complementing field observations of clearcuttings and interviews with implementing actors, the content of these documents was investigated. The analysis of Forestry Agency prescriptions and advice from the entire country also provided comparative information as to variation in environmental protection policy between different counties and districts. This analysis was carried out by Olsson (1987).

All twelve prescriptions were analysed. The NBF requested all County Forestry Agencies to forward every filed "Instruction and Advice" for the use of this research. As a result, 842 instructions and advice corresponding to 35 per cent of the total instructions and advice issued were examined. The remaining documents may have already been destroyed by the CFAs as they are not of further use to the authorities after clearcutting is completed. It is impossible to examine to what extent the content of the investigated instructions and advice is representative of the total amount of written instructions and advice. An additional problem is that much of the Forestry Agency environmental protection advice is orally communicated to the forest owners and forest managers. The share of written and oral advice probably varies in different counties and districts, according to local traditions.

However, there is no reason to assume that the sample forwarded to the National Board of Forestry was biased, other than that some districts and counties were over-represented and others under-represented. It was therefore considered appropriate for the following analysis: (1) what aspects of environmental protection were mentioned; (2) the extent of the protective measures demanded; (3) whether the EPS, the municipality or an environmental organisation had been consulted; (4) the forest owner category to which the document was addressed; (5) which CFA district had issued the document; (6) the age of the Forest Agency officer who signed it [19]; and (7) the date it was issued.

NOTES

[1] Flora, fauna and recreation researchers' recommendations were studied (Ahlén et al., 1979; Ingelög 1981; Hultman 1983) and fauna researchers at the Swedish University of Agricultural Sciences, Faculty of Forestry, Umeå, reviewed the check list.

[2] The categories of environmental measures are shown in Figure 4.1. Six of the forty-four categories on the

check list were, however, never detected in the field. Therefore, Figure 4.1 contains a total of 38 categories.

[3] The content of such plans is described in Chapter 3.

[4] Of practical reasons, only those areas which I myself had investigated could be re-evaluated in this way.

[5] Soil scarification is generally carried out by a machine, which tears away the ground vegetation on small plots, or rows, thus exposing the mineral soil. In these plots, the new seedlings are easily established, because there is less competition from surrounding vegetation and the microclimate in the plots favours their growth.

[6] Protection measures, such as the retention of bogs and fens, steep hillsides and bedrock areas are here labelled "technically and economically motivated" because they are preserved not primarily - or only- for environmental reasons. Rather, they are retained because they are technically difficult to cut, or because the cost of cutting the trees exceeds the revenue.

[7] Ordinary Least Squares.

[8] The highest values were r=0.42 between size of clearcut and private company harvest, and r=0.32 between manual and private individual harvest. Remaining correlations were all below r=0.30.

[9] I kept the minutes of all the discussions. These minutes also used by the national authorities when they formulated the results of the conferences in a document, containing common guide-lines for cooperation between EPS and CFA in forestry environmental protection matters (National Environmental Protection Board 1984, SNV pm 1827).

[10] The three counties were Västerbotten, Västernorrland and Kronoberg; in addition, one EPS officer from Norrbotten was interviewed per telephone.

[11] Practical reasons were, for example, that the person lived too far away (a few private, individual forest owners lived in other parts of the country), that he had moved to another town, or that he was ill (in one case even passed away).

[12] "Forest professionals" here refer to those actors who are directly involved in clearcutting operations. Therefore, private, individual forest owners are also included when the term "forest professionals" is used throughout this thesis, although their profession may not be in forestry.

[13] The classification of respondents as to work tasks are defined in Chapter 3.

[14] When comparing Tables 2.4 and 2.5 the number of Forestry Agency personnel differs with one person. One of the forest owners who was interviewed worked at the Forestry Agency. He was thus employed by the Forestry Agency, although he was interviewed in his role of a forest owner.

[15] Group interviews were carried out fourteen times, with totally 39 people; thus the remaining 76 people, or 2/3 of the interviews were made with one person at a time. Nine of the group interviews were with two persons at a time, two with five (forest workers) at a time, and the rest with three or four people. The majority of group interviews were forest workers who took a lunch or coffee break. (I could not disturbe them too much at work.) Three times were different categories of workers interviewed together: three forest workers were interviewed together with their timber officer and twice the timber officer was heard together with the forester.

[16] For example, when interviews were made with a group of forest workers together, one or two usually discussed more than the others. I perceived the others' consent to the statements made as long as they did not express any other views. Personal data (questions 28-38 in Appendix 2) were, however, always collected separately for each individual. I was aware of the problem with interpersonal relationships within the groups, which may pressure deviants to conform, especially with controversial issues (Compare Kidder and Stewart 1975). Therefore, I tried to register the non-answering person's reaction to the answer.

[17] The tape-recorder was inoperative (without my knowledge) during one day in Östergötland; thus five interviews were not recorded. One of these respondents' answers were noted down during and immediately after the interview (after I discovered the defect), but most of the other respondents' answers were left blank.

[18] The following tables are based on respondent's exact answers: 6.1, 6.5, 6.6, 6.7, 6.8, part of 6.10, and 6.11. Interpretations were made with tables: 5.1, 6.2, 6.3, 6.4, 6.9, part of 6.10, 6.12, 6.13 and 6.14.

[19] The information on year of birth is available in the Forestry Agencies' personnel catalogue.

3 Political/administrative structure and culture of the Swedish forest sector

Forestry in Sweden is a dominant form of land-use, and contributes to a large extent to the Swedish economy. [1] Woodlands total 23.4 million hectares and account for 57 per cent of the land area. Contrary to what many foreigners may expect, the majority of Swedish forests are privately owned. Only 26 per cent of the woodland area is publicly owned. One-fourth of the forests are owned by private companies and almost half of the forests (49 per cent) by private individuals. The forests are accessible to the public, with certain restrictions, according to an old tradition in the Scandinavian countries - the Right of Common Access. [2]

Regardless of forest ownership, Swedish forests are under the jurisdiction of the Forestry Act, which contains general guidelines on forest management, including requirements to protect the environment during forest operations. Setting-aside of larger forest areas for nature conservation purposes is regulated by the Nature Conservancy Act. Two different government agencies with regional branches or sections are formally involved with environmental protection in Swedish forestry. One is the National Environmental Protection Board (NEPB) which defends environmental interests according to the Nature Conservancy Act and the Environmental Protection Act; the other is the National Board of Forestry (NBF), the task of which is to supervise and give advice according to the Forestry Act regulations.

This chapter describes and analyses the forest sector in Sweden, with special regard to environmental issues. It includes an analysis of: (1) environmental authorities and environmental legislation; (2) forest authorities and forest legislation; and (3) forest-ownership and management organisation, with the purpose of examining the implementation of environmental protection in Swedish forestry.

Environmental authorities and environmental legislation

Although the forest authorities hold the main responsibility for achieving environmental policy goals with forestry activities, environmental authorities also often deal with questions that include forestry and the use of forests. Activities undertaken by the National Environmental Protection Board affect forest conditions, for example the acidification of forest soils, and forest death through air pollution. In addition, forestry itself affects the environment, for example through water pollution from the use of chemicals. The Environmental Protection Act regulates industrial and public health policy, and the Nature Conservancy Act includes regulations for national parks, nature reserves, landscape protection areas and protection of rare plant and animal species.

In co-operation with the Environmental Protection Section (EPS) at the County Administration, the NEPB also initiates inventories of environmentally valuable areas. (Recently national inventories of wetlands, virgin forests and mountain forests were completed.) The inventories constitute the basis for Nature Conservation Plans, such as what areas should be protected and purchased by the State to preserve natural conditions. Part of these areas are forests. Priorities as to what areas should be preserved are set by the NEPB together with the EPS at county level. Thereafter, negotiations with the owners of the land to be purchased and protected are carried out by the environmental authorities.

The current state budget for compensating land-owners amounts to 40 million SEK per year. This implies that only a couple of forest areas may be reserved each year in the entire country, depending on their size and timber volume. Forest land is generally expensive and recently almost the whole budget was used for only one area. [3]

Another factor contributing to the low activity on purchase of forest land for nature conservation purposes is the small number of EPS personnel. While environmental protection programmes and tasks have increased enormously during the last ten years, the EPS budgets and personnel situation have not at all responded. The EPS officers are loaded with administrative work, which increases as government programmes are added to provide citizens with acceptable environmental conditions. Few of the EPS officers can be reserved for forest land-use planning and forest protection. Depending on local conditions, there are approximately one or two EPS officers in each county, responsible for all forest matters. Field inventories and negotiations with land owners are often very time-consuming; hence, the preservation of forest areas through public purchase is hindered both by lack of adequate funding and staff.

The EPS is consulted, together with the County Forestry Agency [4] (CFA), if forest operations impact major conservation interests, for example if a forest owner wants to clearcut a virgin forest or an area with great importance to outdoor recreation. Such consultations may be voluntary, with the intention of hearing the EPS' comments on suggested forest operations; they may also be compulsory according to the Nature Conservancy Act if the forest area in question is proposed for a national park or nature reserve according to the Nature Conservation Plan. With some specific forest activities the EPS must be consulted: draining of wetlands, soil scarification of clearcut areas through ploughing (a method used in areas with cold climate) and use of herbicides. In addition, the EPS may demand to be consulted by the County Forestry Agency before a clearcutting is allowed in important recreational areas, areas with cultural artifacts [5], areas with importance to landscape protection, rare plants and/or animals. These areas are often specially marked in the Nature Conservation Plan.

Forest authorities and forest legislation

The National Board of Forestry is located in Jönköping, southwest of Stockholm. Twenty-four regional branches, one in each county, aim at transforming the central directives into practical measures, such as information activities, forest inventories and management plans, and administration of subsidies for road-construction as well as for some silvicultural measures which increase harvest operations and promote forest growth. The County Forestry Agencies are headed by a board, in the same way as the national agency, consisting of politically appointed representatives and personnel representatives. The board decides only in principal matters.

Each CFA is split into local districts, headed by a District Forest officer (DFO). There are about 250 such local districts in the whole country, allowing for a fairly close surveillance and co-operation at the local level with different forest owners and forest organisations. The DFOs are responsible for the daily supervision of the Forestry Act.

Until in July 1980, the County Forestry Agencies were independent, to a certain degree, from the national NBF. The reason is historical - the County Forestry Agencies were initially created at the regional level accompanying the first Forestry Act, which came into practice in 1905. In 1941 the National Board of Forestry was created as an advisory board. A few years later, the NBF obtained the authority to draw up the annual budget for each county board.

In 1980 the CFAs became branches of the NBF. The conviction that success in the work of the CFAs was due to their independent position and local ties explains why CFAs became national government organisations so comparatively late (Stjernquist 1973, pp.95-6). Still today, the work of most County Forestry Agencies and their DFOs could be characterized by close ties between private, individual forest owners and the forest authority. This may partially be explained by the Forestry Agencies' historical role, perceived by themselves as representatives for the collective forestry interests in the counties, rather than public authorities.

In addition to being responsible for the implementation of the Forestry Act requirements, the Forestry Agencies engage in commercial forest activities. They sell seedlings, which are produced in the Forestry Agencies' own nurseries, and they provide labour to carry out planting, clearing and thinning operations, mainly to private forest owners. [6] Other services, which the Forestry Agencies provide include preparing Forest Management Plans for private forest owners and marking trees to be felled with thinning and clear-cutting operations. The above business activities amount to 79 per cent of the total budget. They are financed through fees and income from sales, while only 18 per cent of income comes from the state budget. [7]

The Forestry Agency officers thus have a split role in relation to the forest owners; they must be interested in maximizing profit from NBF forest activities in order to maintain or expand their organisation. At the same time it is their obligation to enforce the Forestry Act. Together with the historical role described above, these commercial activities lead to a general reluctance to affect the forest owner's behaviour through legal punishments. The typical Forestry Agency officer prefers to use persuasion and friendly advice in order to make the forest owner comply with Forestry Act regulations. [8]

The Forestry Act is a fundamental public regulatory framework for the management of Sweden's forests. The act's main purpose is to sustain a high and valuable timber yield from the forests, and in doing so, pay heed to nature conservation and other public interests. It therefore has regulations, applicable to all forest owners, for the establishing of stands, cutting of forests, protecting natural areas and combating insect damage. The chief responsibility for forest policy is vested in the Ministry of Agriculture, whereas the practical application of forest policy rests with the NBF and its County Forestry Agencies. The general political goals of the Forestry Act, which are described above, are defined by the Parliament. These goals are, however, not operative but could rather be characterized as a policy frame for further regulation. The legislators formulate the extent of environmental protection in the Forestry Act: [9]

> The government, or public authority selected by the government, is authorized to announce prescriptions concerning environmental protection measures to be considered in forest management, such as size and design of clearfellings, regeneration, retention of forest areas and extension of forest roads.
>
> The authority given does not include prescriptions announced which intervene to the extent that present land-use (in this case economic forestry) is considerably damaged (emphasis and parenthesis mine). [10]

The goals are ambiguous; moreover, they appear contradictory: the Forestry Act is supposed to promote forestry activities and at the same time protect the environment. The application of the legislation can be adapted according to the decisions of the Forestry Agency and their officers. As will be discussed further, there exists no "definite" judicial interpretation of the trade-off between nature conservation and economic forestry interests.

The regulations are worked out in detail by the NBF and the authorities make interpretation on a case to case basis. According to the decree from the Cabinet, the environmental protection guidelines of the Forestry Act were established by the NBF after consultation with the National Environmental Protection Board. Conflicts between economic forestry and nature conservation interests remain according to statements in the Forestry Act. They are presumed to be solved through the remiss process before the legislation is enacted and, afterwards, through negotiations between NBF and NEPB representatives in detailed regulations.

There are several explanations why legislators give the NBF such broad discretion. Maybe, the intention was to avoid having to make actual policy choices, thereby delegating to an administrative agency the often difficult and delicate decisions. Beyond that, though, is the explanation that it is impossible for a legislature to establish detailed technical decisions on thousands of specific issues (Kelman 1981). Discretion is here defined as "the form of decision-making freedom acquired and defended by peripheral agencies of government" (Ham and Hill 1984, p.148). Studying "discretionary decision-making requires a consideration of social processes internal to the organisation and a study of the attitudes and beliefs of those who have to interpret the rules" (Ibid., p.163). The concern of discretion is also normative: "under what circumstances may discretion be said to be a problem, and for whom?" (Ibid., p.173).

Both enforcable prescriptions and general recommendations are included in the regulations. Most of the details are described in the general recommendations, while the prescriptions tend to be more vague. Although the regulations issued by the NBF are detailed as to examples of specific features to be protected, they are vague as to the extent to which these features should be protected. To give an idea of the content of the regulation, an example is reproduced below: [11]

Prescription:

> Nesting trees for rare and threatened species shall be retained along with protective trees and other vegetation around the nest. Living and dead trees and bushes with special importance to the survival of animals shall be retained <u>to an appropriate extent</u>. Other small environmentally important areas to flora and fauna shall be protected in the same way (emphasis mine).

General advice:

> "Nesting trees" imply trees with holes, trees with large nests and snags; also such trees, single or groups, which have special qualifications for developing into valuable nesting trees... To protect eagles and eagle owls the protective zones which may be required by the Forestry Act are not always sufficient. Usually the forest must be retained in a circle of 200 metres aroung the nest. The same is applicable to certain other birds of prey... Trees with berries and certain other nutritive trees and bushes, for example apple, cherry, whitebeam, hazelnut, juniper trees, blackthorn and dog-rose bushes, should be retained. Living and dead old broad-leaved trees, preferably in groups; snags and single dead conifer trees; single large, old wind-throws may be retained for nutritive purposes and nesting trees for bats, wood-peckers, other hole-building species, rare insects etc. These measures are also of importance to flora: lichens, mosses and mushrooms...

Environmental protection should be considered "to an appropriate extent". What is "appropriate" is then left to interpretation and could be discussed from case to case by the DFOs. The general advice provides some guidance, but the substantive content is shaped by what aspects the DFOs emphasize or restrain from emphasizing during their work. Hence, much of the agency policy is actually decided at the lowest level, that is by the officers in closest contact with forest owners and forest operative personnel.

Since the year of 1980 the forest owners are required by the Forestry Act to prepare a Forest Management Plan. [12] The plan should include description of forest conditions as well as suggestions for management and be financed by the forest owner. Often, the County Forestry Agency prepares

Forest Management Plans (but the forest owner pays for it) while carrying out forest inventories. Such forest inventories are recently completed for approximately one-third of private, non-industrial forest land and is financed by the State. [13]

If forest or environmental authorities or municipalities consider the forest area especially valuable to recreation, landscape protection or protection of rare species, the Forest Management Plan may be extended to include environmental impact assessment. The plan is then usually financed by the Forestry Agency. However, only a few forest areas are inventoried with such special emphasis on environmental protection. [14] The majority of Forest Management Plans do not include suggestions for environmental protection. [15]

Forestry environmental regulation background

The development of environmental legislation in forestry officially started in 1972, when a team was established at the Department of Agriculture to investigate the extent and effects of clearcuttings. According to their report, clearcutting is the preferred harvest method on the majority of forest land. They also had suggestions for mitigating environmental damage:

> The negative impression from clearcutting could quite easily be diminished... Such measures as to retain single trees and buffer strips adjacent to residential areas and along roads and lakes, to adjust the geometry of the clearcut area to landscape features and to clean old forest paths from logging debris... should all be possible to combine with economic forestry... In order to express this common desire, the requirements of the Forestry Act should include a general deference to environmental protection. [16]

An extensive remiss process preceded these suggestions which were not challenged by any of the political parties. On the contrary, both the remiss statements and the discussion in Parliament showed great unanimity on the need of environmental protection with all forest activities. In 1974 the Forestry Act was altered in two aspects relevant to environmental protection: a general recommendation to the forest owners to pay heed to environmental interests (§ 1) and a requirement to notify the Forestry Agency before clearfelling a forest area (§ 8a, in 1979 changed to § 17). Through the advance notification it became possible for the Forestry Agency to recommend certain environmental protection measures before the harvest operation commenced. [17] Several of the contributions to the debate in Parliament pointed out such a motive for the notification duty. [18] The statements in Parliament were, however, not entirely positive; the three non-Socialist parties all had objections, although not to the environmental aspects:

> A common notification requirement... would imply considerable inconvenience for individual forest owners, while the work load of County Forestry Agencies would greatly increase... We cannot accept that the CFA resources aimed for forest improvement measures become more restricted, especially since corresponding environmental gains will not be obtained (Conservative Party Bill 1974:2005).

> A notification duty according to the suggestion from the Minister of Agriculture without being accompanied by additional personnel resources to the CFAs would lead to a reduction in forest regeneration efforts (Centre Party Bill 1974:2008).

> Our opinion is that particularly from a nature conservation standpoint it would be more appropriate to concentrate notification requirements to such areas which will be classified as conflict areas... If the County Administration decides and communicates after the inventory which areas will be subject to consultation according to the Nature Conservance Act, there will be no need for notification to the CFA because of environmental protection reasons. Contacts may then be directly established between the CFA and the EPS (Liberal Party Bill 1974:2011).

As long as environmental protection issues are held at a general, abstract level, all political parties and different interest groups agree. If, however, the trade-off between environmental protection and economic forestry has to be specified more precisely, different standpoints come forward. The core of conflict is the interpretation of how much protection could be required without "considerable damage to economic forestry". If this limit is exceeded, the forest owner must be economically compensated and the area protected through the Nature Conservancy Act.

The expressions formulated in the legislation and the statements made by the Minister of Agriculture during the preparatory speech in Parliament were considered by the environmental authorities to be insufficient and partially contradictory. Therefore, in 1976, the National Environmental Protection Board set up an informal committee, the "SKONA-group" [19], consisting of representatives from environmental and forest national authorities along with industrial and non-industrial forest owners. The group was supposed to have an advisory role to the NEPB in the preparation of guidelines for applying the Nature Conservancy Act in forestry. The aim was to establish a definition of the wording of the act concerning the limit for economic compensation to the forest owner - specifically, the interpretation of "considerable damage to economic forestry".

A major conflict appeared in the SKONA-group between the environmental protection authorities on the one hand and forestry representatives on the other. The environmental representatives argued that the "economic damage" should be related to the total value of the real estate. Large-forest owners should be able to accept more protected areas than small-forest owners without being economically compensated. The forestry representatives preferred to relate the "economic damage" only to the specific area affected, for example the clearfelling in question. Personal economic damage should also be considered. Consequently, the environmental protection interests should be equally judged independently of size of forest-ownership. Every square metre retained for environmental protection purposes should, with the forestry representatives view, be valued the same.

The SKONA-group was dissolved without reaching an agreement. Westerlund (1980) scrutinizes in a judicial analysis the Nature Conservancy Act rules regarding compensating land-owners without being able to establish the limit for economic compensation. As he concludes in his last lines: "We must await litigation in courts". So far, very few cases have been taken to court [20]; none of which have involved the environmental protection regulations of the Forestry Act. The extent of environmental protection required by the Forestry Act therefore remains in dispute.

At the alteration of the entire Forestry Act in 1979 (the previous dates from 1948), the environmental protection recommendations in the § 1 were moved into a paragraph of their own - § 21. The reason was the amendment of a penalty clause connected to the environmental requirements. This was motivated by a Royal Commission:

> In order to assure compliance with the Forestry Act regulations for environmental protection they should be combined with sanctions in some form... Inevitably, both the forest owners and the government authorities will, at least in some cases, face difficult situations if the prescriptions could be generally enforced by penal rules. In certain situations, however, violations should be punished... Primarily, it should be considered if the desired result may be obtained through information and advise, like the rest of the Forestry Act supervision. [21]

The Minister of Agriculture as well as all political parties, except the Conservatives, supported this suggestion. [22] The Conservative Party was afraid the legal requirements would be pushed too far, and that forest owners would not be adequately compensated. [23] The final decision in Parliament corresponded to the proposal made by the minister:

> ...that the prescriptions issued within the Forestry Act generally should not be associated with penal rules. Instead, the County Forestry Agency should decide in which situations it is necessary to communicate a specific prescription, which if neglected may lead to a fine. [24]

No pecuniary penalties can therefore be imposed if the Forestry Agency has not prescribed what environmental measures must be fulfilled. The public, for example environmental organisations, cannot sue a forest owner who does not comply with the regulation; all they can do is to lobby the government authorities to react, or to attract attention through mass media. Most environmental features, such as nesting trees, buffer zones etc. must be recognized before harvest commences, if they are to be protected by a prescription issued by the Forestry Agency. As will be discussed below, very few cases have attracted that much attention.

Resources for monitoring

As mentioned, compliance to the Forestry Act regulations is monitored by a notification system of clearcuttings. Other silvicultural treatments, such as clearing, spacing, thinning and road-building are carried out without application [25] or permission from government agencies. It is therefore difficult for the Forestry Agency officers to monitor environmental impacts. Clearcutting is regulated in detail with regard to stand age, allowed proportion of clearcut area compared to the total forest holding, and regeneration measures. Every clearcut larger than 0.5 hectare must be reported at least one month in advance to the County Forestry Agency. [26] If the proposed clearcut is located within an environmentally valuable area identified by the EPS, notification must be made two months in advance to allow for review by the EPS.

Clearcut notifications are made to the County Forestry Agency directly, where they become official documents, available to the public (according to the Swedish principle of public access to official records). Such notifications contain information on the forest owner and the size and location of the area proposed for clearcutting as well as regeneration methods. Information on environmental values are, in general, absent or meagre. After being recorded at the county office, the notification is channeled to the District Forest officer. If required, the notification is also submitted to the EPS at the County Administration.

Administrative routines vary slightly between different counties. In some counties the Environmental Protection Sections provide the DFOs' with all information (including maps) of environmentally valuable forest land. The DFO may then decide himself [27] whether or not to contact the EPS.

In other counties, the decision remains with the EPS to apply the Nature Conservancy Act in environmentally valuable areas. Informal and personal contacts between the EPS officers and the County Forestry Agency are frequent.

Relations between environmental and forest agencies vary in the counties investigated in this study. For example, in Västerbotten the responsibility for monitoring environmental protection rests with the County Forestry Agency, except for one river valley where forest owners have to apply to the Environmental Protection Section of the County Administration before clearcutting.

In Västernorrland the responsibility is split between the CFA and the EPS. Before 1980 the EPS investigated the environmental impact of proposed clearcuts within environmentally valuable areas. This work was quite extensive, since approximately 11 per cent of the county required consultation with the EPS. Every year almost 400 clearcut notifications in Västernorrland county were passed on to the EPS. [28] After 1980 the responsibility for all these areas was passed on to the County Forestry Agency, along with EPS maps, inventories and data regarding important environmental values on forest land.

In Östergötland the responsibility is also split between the authorities as outlined by the Nature Conservation Plan, which describes major environmental interests affecting land-use management. The Forestry Agency officers are to discuss requirements for environmental protection with the EPS officers when clearfellings affect areas described in the conservation plan.

If the forest area proposed for clearcutting is important to local recreation or nature preservation interests, the Forestry Act suggests the CFA consults the municipality. According to the interviews with Forestry Agency officers, only three out of twenty-four said they ever had consulted municipality representatives. The contacts between municipalities and CFAs are, so far, essentially "experimental" and general routines not yet established. According to the recent Building Act of 1987, the municipalities are required to draft Master Plans, which include land-use planning on forest land. This planning is to be made in consultation with affected forest and environmental interests. Previous planning efforts have also implied some contacts between the municipalities and forest authorities. As recreation and conservation interests grow, the pressure is growing also from government authorities to work out formal methods for such contacts.

Many clearcuttings are initiated by the DFO, who advises the forest owner to harvest the mature forest. It also happens (according to one DFO this is a rule in his district rather than an exception) that a clearfelling begins without the notification being made one month in advance. The DFO may give permission to continue a clearcutting once it has started, if it is legal in all other aspects.

The DFO may react to the proposed forest cutting in different ways: he can prohibit clearcutting; he can prescribe specific measures that could be legally enforced; or he can give specific advice, with the status of a recommendation. Forest conditions within a district are often known to the DFO both through his field experience and through Forest Management Plans which are kept at the district office.

Every District Forest officer receives 100-300 clearcut notifications annually; therefore priorities must be made. According to the interviews with District Forest officers, their motivation for monitoring forest owners is to investigate silvicultural practices rather than to check-up on environmental matters. The DFO primarily checks the age of the stand to be clearcut, since cuttings must exceed a certain age limit according to the NBF regulations in the Forestry Act. In general, only twenty to seventy per cent of all clearcuttings are field investigated in advance by the CFA. Most notifications arrive to the CFA in autumn, which makes it impractical to investigate all clearcuts in the field. In cases where the officer has worked with a district for many years, he knows which forest owners are resistent to advice and devotes extra work to monitor them.

Most DFOs say they give priority to investigating non-industrial forest owners over industrial forest owners. The motive mentioned by several of the officers is that the large companies should have enough educated personnel to manage their forest properly.

State-owned forests, managed by the Swedish Forest Service, were not under the jurisdiction of the Forestry Act until in 1980; thus, Forestry Agency monitoring of state-owned forests is newly introduced. [29] There is, however, a tendency expressed during the interviews, that contacts with the large-forest owners are becoming more frequent. Directives from the NBF encourage this development.

Environmental protection measures may be pointed out by the DFO when he receives the clearcut notification. It is important that he recognizes environmental problems <u>before</u> the cutting commences; afterwards, it may be too late, since most environmental values can be destroyed through clearcutting. The DFO may advise the forest owner orally, for example during a field inspection on the area. He may also write an "Instruction and Advice" on environmental protection according to a standardized form provided by the Forestry Agency. The aim is then to give the advice a more formal status, since there is no way a forest owner could be punished if he does not follow the advice. It also <u>documents</u>

the advice, so that the DFO may refer to it and prove he has tried to influence the landowner.

The next step is to write a "Prescription" or "Prohibition", which, if not observed, may result in a fine. From the beginning of 1980, when § 21 was enacted, to the end of 1986 only twelve such prescriptions and prohibitions were issued nationwide. [30] The written instructions and advice are more common and total approximately six hundred annually [31], which implies an average of around two per year and District Forest officer.

Forest owner and management organisation

Most of the public forests are owned by the State (the Swedish Forest Service) and mainly located in the northern parts of the country. The Forest Service is an autonomous trading agency administering the State's woodland and farming holdings. Forest Service operations are to be run along commercial lines, but with great attention paid to environmental protection. The rest of the publicly owned forests belong to the municipalities, the Church or are so called common forests, reflecting an old tradition originating from the farmers' common need for pasture land and domestic cuts of timber.

Private company forests (dominated by two very large companies) are mainly located in the centre of Sweden. Private ownership by individuals is predominant in southern Sweden, but spread all over the country, and distributed among some 240,000 holdings. One-third of individual forest owners do not live on their holdings. [32] To aid private forest owners, there are twelve regionally based Forest Owners' Associations who assist in forest management and co-ordinate timber trade. Their 80,000 members represent almost half of the total forest land owned by private individuals. There is also the Forestry Society, a private organisation offering forest management services to municipalities, foundations and private forest owners.

Forest owners may be classified along at least two dimensions: (1) as public or private and as (2) large- or small-scale. There are great similarities between public and private forest owner organisational settings, whereas organisations of large- and small-forest owners vary to a greater extent. Large-forest owners, such as private companies and the state-owned Forest Service employ their own forest management staff. Head-quarters set up the company goals, calculate their long-term as well as yearly timber yield figures and provide general management plans. On the district level, "foresters" supervise operational planning.

"Foresters", according to the definition used in this study, include forest officers and forest managers who carry out the master planning of forest operations and who are responsible for the management of a forest district. One forester may have one or many "timber officers" helping him with the detailed planning. Depending on where he is employed, the forester may be responsible for a district within; (1) the Forest Service (equivalent to Forest Rangers in the United States); (2) a private forest company; (3) the Forest Owners' Association or (4) the Forestry Society.

Timber officers make the practical arrangements for harvesting and other forest operations, including additional selection from the forester's master plan of forest areas to be treated. "Timber officers" are here defined as all foremen, or low-level supervisors, who are directly responsible for the planning and performance of the operation. They are often former forest workers with vast practical experience in forest management. In the analysis of interview data, "planners" are also added to the group of timber officers. Like the timber officers, planners are working close to the forest workers since they are planning all the details of the harvest operation.

"Forest workers" are here referred to as the ones who actually do the logging, that is cut and transport the trees to the landing. When they arrive, the detailed planning of the harvest is already completed. "Forest worker" is used in the analysis as an expression for both the labourers who use chain saws and the ones who drive forest machines. Sometimes these two groups are divided into "manual" and "machine-working" for comparative reasons. Older forest workers are often self-taught through a long work experience, while the younger most often have attended the forestry line of upper secondary schools.

Small-forest owners - which may be either small municipalities, small private companies and private foundations or individual forest owners - do not usually employ their own management staff. Some private forest owners carry out silvicultural treatment by themselves, which is most frequent among those who live on their holdings and are not employed elsewhere. The smaller the holding, the more the activities are carried out by the owner (Lönnstedt 1974). The majority of small-forest owners hire contractors for the job. This may be effected in a variety of different ways. The forest owner may turn to the Forest Owners' Association, the Forestry Society or other private companies, for example saw-mills, who then take over the operational planning of forestry.

These three types of organisations have foresters, timber officers and forest workers to carry out the logging operation. Their economic offer to the forest owner is calculated from the value of standing timber reduced by the costs for logging. The forest owner may also let somebody mark the stand of timber for felling and present a list of marked trees, from which the harvest yield can be calculated. This marking may be conducted by a private forest-educated person or by some forest organisation, for example the Forest Owners' Association (usually they are then also buying the stumpage sale) or the County Forestry Agency.

On the basis of the list of marked trees, bids are given from different buyers and the forest owner can choose the most favourable offer. The trees for felling are already selected; the price given thus includes the right to harvest all marked trees. All the above forest organisations may bid; in addition, private individuals may offer to carry out the operation. Consequently, a large number of people are involved with harvest planning and operation. Only the forest owner is legally responsible for Forestry Act compliance and is the only person who could be convicted, or fined, if he fails to meet the requirements.

Administrative setting varies between different employers, but two main types could be discerned: (1) foresters and timber officers are responsible for all activities within their district or (2) foresters and timber officers are responsible for specific activities, for example purchase of timber, planning of logging or regenerative measures. Most forest companies, however, combine the two models of competence; the local forester and timber officer assumes all operational tasks but central or regional specialists may assist with guide-lines and instructions.

None of the forest organisations employ specialists on environmental protection and nature conservation. Forestry personnel (i.e. all the categories of workers mentioned above) have a degree in forestry which does not include specialist knowledge on environmental protection. No biologists, ecologists or landscape architects are employed. The same is true for the Forestry Agency personnel. [33] Therefore, forestry administration can be regarded as a self-regulating core of professionals which has its own distinctive standards. Administrators share a professional code of ethics which guides their behaviour (Self 1972, p.226). Differences in attitudes and behaviour are minimized through a homogenized professional staff.

Means of homogenization pointed out by Kaufman (1960) in his study of the U.S. Forest Service are highly applicable also to Swedish forest organisations: (1) the use of professional foresters in all types of specialized jobs; (2) in-service indoctrination and training and (3) rotation of professional personnel by lateral transfer and by diagonal promotion. As a result, local forest personnel are systematically selected and trained for acceptance of policies enunciated by the higher levels (Kaufman 1960).

The homogenization leads to friendly relationships between the Forestry Agency officers and the forest owners and their employees - for example, personnel is frequently transferred between the controlling forest authority and its objects of control. Their common educational background and acquired attitudes towards economic forestry favour the emphasis on timber production as opposed to environmental interests, which will be further analysed in Chapter 6.

NOTES

[1] When no specific source is referred to, the information presented in this chapter is based on official documents and minutes available at the different agencies, and from statistical data from the National Board of Forestry. In addition, the analysis in this chapter is, to a great extent, based on my interviews. In the section "Resources for Monitoring", much of the data comes from the conferences with forest and environmental authorities, and the last section is primarily based on interviews with forest owners and managers.

[2] The Right of Common Access is applicable in Sweden, Finland and Norway - however not in Denmark.

[3] This happened in 1987 when Tandövala (approximately 4,000 hectares) was protected (Minutes from the National Environmental Protection Board, 1987).

[4] The County Forestry Agency is here referred to as the government forest administration. They refer to themselves as "The County Board of Forestry", but as earlier described, they are actually an agency which is headed by a laymen board.

[5] Cultural artifacts are protected through the Ancient Monuments Act, a legislation which prohibits the landowner from destroying cultural resources, such as old settlements, graves and pits for hunting moose. Anyone who wants to operate in such an area, for example construct a road, must pay for the excavation of the ancient monument. Consequently, all ancient monuments and cultural resources belong to the State; they are strongly protected compared to natural features, such as water and plant and animal species. Preservation of ancient monuments is supervised by the National Board of Antiquities and their county level officers. The Regional Inspector of Antiquities works within the County Administration, analogous to the Environmental Protection Section officers.

[6] The term "private forest owner" is used instead of the more descriptive term "private, <u>individual</u> forest owner". "Private forest owners" do not include private

forest companies. When these are referred to, the terms "industrial" and "non-industrial" private owners are used.

[7] The Forestry Agency Organisation 1985-1986. National Board of Forestry, 1987.

[8] Compare Chapter 7.

[9] Forestry Act § 21, SFS 1979:429.

[10] All citations from Swedish official documents are my own translations.

[11] National Board of Forestry regulations 1983:2 to the Forestry Act § 21.

[12] The Forest Management Plan is required, according to the NBF regulations, to be completed before 1993, and shall be renewed every tenth year. Forest owners with small holdings need only prepare a list of appropriate forest operations.

[13] The Forestry Agency Organisation 1985-1986. National Board of Forestry, 1987.

[14] During 1975-1985 Forest-Environmental Management Plans were prepared for a total of 125 areas (125,000 hectares). (The Forestry Agency Organisation. National Board of Forestry, 1987.)

[15] In both the Forestry Agency's and the forest companies' routines for preparing forest management plans there is a code for environmental features, for example the occurance of rare plants and animals. However, according to the interviews, environmental features are rarely inventoried and therefore very seldom appear in the Forest Management Plans.

[16] The Swedish Government Official Report "Clearfellings" by the Ministry of Agriculture: Ds Jo 1974:2, pp.247-8.

[17] There were, however, divergent opinions on whether notification should be made before or after clear-cutting. Some remiss statements (for example the National Board of Forestry, but only a minority of the County Forestry Agencies) advocated notification during or after clearfelling is completed. These issues were debated in the Standing Committee on Agriculture (JoU 1974:52).

[18] Debate in Parliament 1974:145.

[19] The name "SKONA" refers to the Swedish abbreviation of forestry (SKOgsbruk) and environmental protection

(NAturvård). The analysis of the SKONA-group is based on internal documents, filed at the NEPB.

[20] The cases referred to here are cases affecting forestry and brought up according to the Nature Conservancy Act: Two stumpage sales appealed to the Supreme Court (HD verdict no T61/77 and T94/77 and T94/77) considering compensation for retained timber because of landscape protection (Compare Ovegård and Rydström 1982). There was also one case of clear-cutting on small islands where environmental protection measures were prescribed by the EPS (appealed to "Kammarrätten" in Jönköping verdict no 4031-1979). One appeal to the Government (Ministry of Agriculture) questioned the right to harvest a forest area with importance to reindeer husbandry (Government decision no 31/85). Two more appeals to the National Board of Forestry (no 963/84 and 1296/84) concerning the limit for compensation when a forested bog area should be retained according to the Forestry Act because of environmental protection. This case involved several forest owners, who pleaded for a smaller area retained than prescribed by the County Forestry Agency. The NBF decided to meet the demands of the forest owners. (Documents filed at the National Board of Forestry).

[21] Royal Commission Report 1978:6, p.130.

[22] Debate in the Standing Committee on Agriculture JoU 1978/79:30.

[23] Debate in Parliament 1978/79:154.

[24] Government Bill 1978/79:110, p.53.

[25] Road-building is, however, subsidized by the State under certain conditions. Forest owners apply for such grants to the Forestry Agency, which may influence the location and building of the road. Also, in areas specified by the Forestry Agency as having difficult regeneration prospects, the forest owners must apply for both thinning and clearcutting operations. This enables the Forestry Agency to further control the cutting of sensitive forests.

[26] Except in sensitive areas, see above, where application must be made.

[27] "Himself" is used because there are very few women in Swedish forestry, and none of the persons interviewed were women.

[28] Interview with officers at the Environmental Protection Section in Västernorrland.

[29] Church-owned forests are also regulated by the Forestry Act only since 1980.

[30] According to official statistics from the National Board of Forestry, only nine prescriptions and prohibitions were issued from 1980, when the penalty clause was amended, to the end of 1986. Olsson (1987) has, however, found twelve such prescriptions and prohibitions issued during the same time. Ten of these were prescriptions and two were prohibitions. One more prescription had been issued in the beginning of 1987.

[31] According to statistics from the National Board of Forestry a total of 2,424 written instructions and advice were issued during the years 1980-1986. In 1985 and 1986 the number was 645 and 615 documents respectively (Compare Chapter 7; Table 7.1).

[32] Official Statistics of Sweden, Statistical Reports: SOS J 1979:14.

[33] There is, to my knowledge, at least one exception: the NBF recently employed a biologist to answer for environmental policy issues. Although there may be other exceptions, the tendency is clear: forest-educated personnel is preferred.

4 Environmental protection in clearcuttings

Examination of policy outcome consists of field investigation of clearcuts, where environmental protection measures are evaluated in relation to the Forestry Act requirements as well as researchers' preservation recommendations. [1] The dual aim with the field inventory is to establish a measurement on the extent of environmental protection considered in practice, and to analyse what factors contribute to the outcome. Hence, this chapter not only describes the environmental protection outcome with clearcuts in practice, but also examines an explanatory model in relation to the clearcut areas investigated. The method of analysis used in the field investigation is, however, limited to those factors that could be measured without interviews. It starts with a measurement of the policy outcome and identifies common factors which may contribute to "successes" or "failures" related to the programme.

In a multiple regression explanatory model [2] the degree of environmental protection is analysed, depending on: (1) what harvest operation system is used; (2) who has fulfilled the operation (e.g. public or private organisation); (3) in what county and (4) the size of the clearcut. The four different factors are tested as expressions, or underlying variables, for the variables earlier outlined (See Figure 1.2). The factors in the model and the variables in the framework for implementation analysis are assumed to be related as follows:

"Harvest operation system" in the model is a direct measure of "harvest technology", but also, indirectly, of "forest economic considerations". However, as mentioned in the introduction, forest economic considerations are defined in the framework for analysis as those economic considerations which remain when a certain harvest technology is selected. Harvest technology may affect the degree of environmental protection both by technical adaptability to different protective measures and by the scale of treatment. Large machinery, such as harvesters and processors, may be used in a more stereotyped way compared to when the cutting is performed manually with chain-saws. High technology may, however, not automatically lead to stereotyped behaviour if the machines are well adapted to, for example, small-scale considerations. There might also be differences between the harvest operation systems because "implementing actors' attitudes and knowledge" may vary between, for example, manual workers and operators of large forest machines. The hypothesis would be that the manual workers are more interested in environmental protection through their "closer contact" with Nature compared to the forest machine operators.

"Who has fulfilled the operation" is assumed to be related to "forest-administrative culture and structure", to "implementing actors' attitudes and knowledge" and to "forest economic considerations", since different forest owners and managers may have different forest management aims as well as different organisational settings. For example, private interests may be more inclined to emphasize monetary values than would public agencies. The Forest Service, being a state agency, may thus be expected to pay more attention to environmental aspects than would most private forest owners and managers. Different forest owners and managers may also emphasize environmental protection to various extent, which may lead to divergent attitudes and knowledge among their personnel.

The factor "county" is examined in the model because there may be different traditions in "forest and environmental administrative culture and structure" as well as different "implementing actors' attitudes and knowledge" in the counties. Differences in forest ownerships and administration between the counties might thus give rise to geographical variations. The counties differ slightly, as previously described, in terms of collaboration and division of labour between the environmental and forest authorities. They may also have emphasized environmental protection to various extent, and thereby created different attitudes among implementing actors.

It is further assumed that "size of clearcut" is related to "forest economic considerations" as well as to "harvest technology", because highly mechanized cuttings and great demands for economic profit may lead to larger clearcuts. Possibly, there also exists an independent effect of size of clearcut area on the degree of environmental protection. Such an effect may be more stereotyped planning and execution of a large harvest operation when the clearcut is large, as compared to a small-size cutting. According to the Forestry Act, the aim is to limit the size of each clearcut. In forest areas intensively used for recreation the clearcut is recommended by the NBF to not exceed five hectares. The average size of Swedish clearcut areas amounts to 6.3 hectares, according to the notifications made to the Forestry Agency [3], but there are great variations between the forest owner categories (and between different parts of Sweden). State and private company-owned forest account for the largest clearfellings - 50 per cent of their total clearcut area consists of clearcuts larger than 20 hectares each. Non-industrial private forest owners have only 8 per cent of their total clearcut area in this size category. There is no indication that the forest owners have been influenced to reduce these sizes as recommended by the Forestry Act in 1974 (Eckerberg 1982).

Environmental protection results

The average clearcut of the investigation was 16.6 hectares. On the average, there were nine different environmental protection measures to consider. The three most common features were old, dead conifer trees, trees with berries, and trees with holes used as dens by different animals. Before harvest there was no significant difference in amount of environmental protection detected between different sizes of clearcuts. This is due to the method of inventory. The environmental protection measures only register whether or not the specific feature is present. The extent of the feature is thus not measured. [4] Hence, seven categories of environmental protection measures were to be considered at the average small-scale cutting (4.7 hectares), while large-scale forestry with average clearcuts of 30 hectares had ten categories of environmental protection measures to consider.

The total results of all environmental protection measures are described in Figure 4.1. Each row represents the frequency of a specific category of environmental protection measure. According to the regulation of the Forestry Act a new clearcut should not adjoin another clearcut. Of the investigated clearcuts, 83 per cent were not adjacent to old clearcut areas. With the remaining areas, which bordered another clearcut, a buffer between the new and old clearcut was left in approximately half the cases.

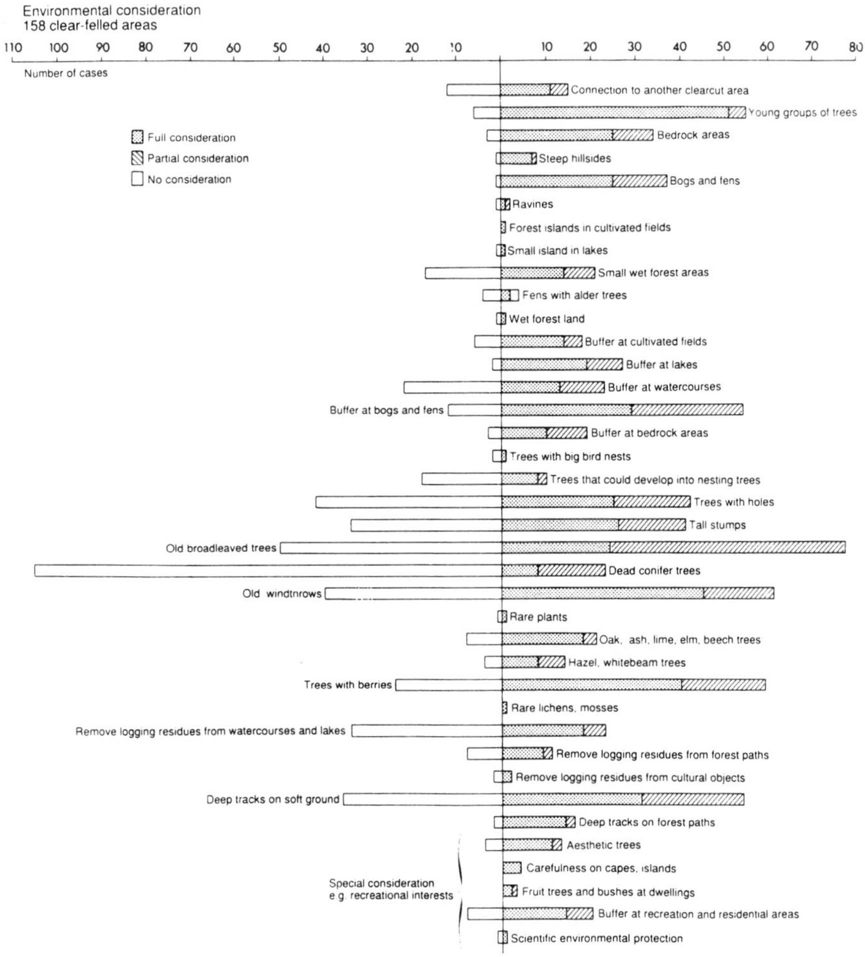

Figure 4.1 Degree of environmental protection achieved by different measures

Technical/economic and aesthetic aspects

There was great variation in the degree of environmental protection amongst different categories of environmental protection measures. Categorizing these into three environmental protection aspects: (1) technical/economic, (2) aesthetic, and (3) floristic/faunistic - reveals clear tendencies (See Figure 4.2). When environmental protection measures are technically and/or economically favoured - such as retention of young stands of trees [5], bogs and fens, bedrock areas and steep hill sides - the compliance with the environmental regulation is very high. Forest operators do not risk losing equipment in a bog, unless it is frozen. It is also quite difficult to harvest trees on steep slopes or ledges. Usually, the economic gain of clearcutting bedrock areas or bogs and fens is very small. In these cases, environmental protection interests often work hand in hand with economic interests, but environmental considerations are not the primary reasons for protecting the areas.

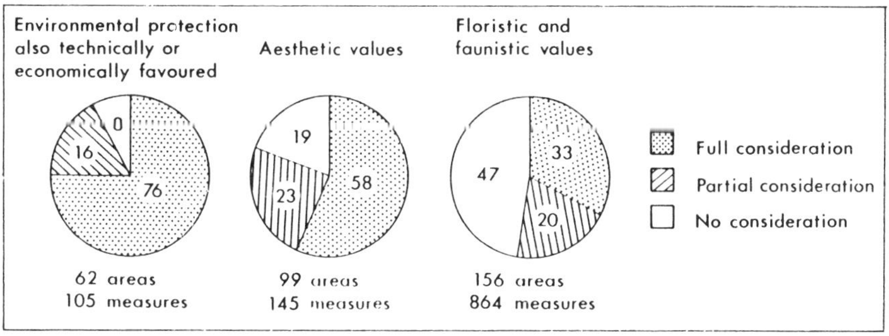

Figure 4.2 Degree of consideration to different environmental aspects (Per cent of the total number of environmental measures required)

Aesthetic aspects were also often cared for. These aspects were emphasized by the National Board of Forestry information concerning environmental protection in 1974. Along lakes and cultivated areas the retention of vegetation mitigates the visual impact of the clearcut. Consideration to recreational interests, or because of adjacent dwellings, was required in 22 per cent of the investigated areas. Trees at halting-places and forested capes at fishing lakes were most often saved. Near summer dwellings fruit trees and bushes were retained. The eight cases where a buffer should have been left adjacent to residential and recreation areas but "no consideration" was shown, are evaluated based on the presumption that at least some vegetation should have been saved. However, the people living in these summer cottages might have requested the surrounding trees harvested in order to let in the sun.

Flora and fauna aspects

Floristic and faunistic environmental protection features were identified in 156 of the total 158 clearcuts, yet only half of the 864 measures investigated were protected. However, some of these features were very rare, such as ravines, forest areas in cultivated fields, small forested islands, wet forest land, trees with nests for birds of prey and rare plants (which were only present at three, or less than three, clearcuts). Scarcity made conclusions inadequate on these and other objects of environmental protection, for example fens with alder trees (Alnus sp).

Other features, however, were more common and therefore easier to evaluate. Forest buffers along watercourses, lakes, bogs, fens, and bedrock areas most often comprised a variety of bushes and trees, making buffers valuable to plant diversity and animal habitat. When the clearcut bordered a lake or a cultivated field, a buffer zone was usually retained, but little vegetation was left when the buffer zone was located in the middle of the forest as was the case with buffers along bogs and fens, watercourses and bedrock areas. The high frequency of "partial consideration" to buffer zones was due to partial harvesting along the borders, therefore disrupting the buffers' variation in heights and species. Borders were quite frequently thinned, leaving mostly the tall conifer species, especially the Scots pine (Pinus sylvestris). The clearing of small trees and bushes on the clearcut area was usually carried out before the harvest operation if mechanized equipment was used. When the cutting was manually performed the clearing was most often carried out during the harvest.

Small, wet forest areas and fens with alder trees were saved to "full" or "partial" extent with only half the cases. The species of plants and animals which inhabit, or depend on, wet forest areas, were therefore threatened by economic forest management. In addition, deep tracks from forest machines were more noticeable in wet forest areas.

Many of the endangered species depend on decayed wood or dying and dead trees, called "snags". In the virgin forest there is an abundance of wood in different stages of decomposition, while economic forestry aims at removing snags. Dead trees were only saved in 18 per cent of the areas where such trees had been observed before the cutting commenced. Old windthrows were left to some extent in 60 per cent of the areas. All these trees had been dead more than two years, where there was no longer a risk for noxious insects (e.g. Ips typographus). Windthrows were saved more often than standing trees, probably because the windthrow was more difficult to cut and was more often decayed. In addition, when the harvest was carried out during the winter, the windthrows were hidden by the snow.

Sometimes snags were cut because they were obstacles or work hazards during harvesting. The problem was prevalent when it came to saving tall stumps and trees with holes. In the investigation, "tall stumps" were referred as dead trees without a top and with diameters exceeding 20 centimetres (DBH) [6] - potential nesting trees for example for the ural owl (Strix uralensis). "Trees with holes" referred to all trees, both dead and alive, with a den. Many were used by wood-peckers. The category "trees that could develop into nesting trees" did not include tall stumps but solely old pine trees with flat crowns and branches, strong enough to hold a nest for birds of prey. Half of the above nesting trees (including potential nesting trees) which were registered before harvest were felled during the harvest operation.

Old broadleaved trees, such as sallow (Salix sp), aspen (Populus tremula), alder (Alnus sp), and birch (Betula sp) were found at 80 per cent of the areas. Some of them were already dead or dying. "No consideration" was shown in 40 per cent of these areas and "full consideration" in 20 per cent. Usually, "partial" protection referred to a single tree retained on the clearcut. Groups of broadleaved trees were seldom retained. Oak (Quercus robur) and ash (Fraxinus excelsior) were found in the county of Östergötland. They were retained to "full" extent in 60 per cent of the areas; however, usually only a single tree was left. In the county of Östergötland there was also some hazel (Corylus avellana) and whitebeam (Sorbus aria and S. intermedia), which were retained to 80 per cent. There is a severe threat to deciduous forests in Sweden, because old pasture land and small areas of broadleaved trees are often converted to softwood plantations. It was, however, not possible to evaluate the magnitude of site conversion in this investigation.

Trees with berries are of high food-value to many animal species. Rowan (Sorbus aucuparia) and juniper (Juniperus communis) were observed at more than half of the areas before cutting. "No consideration" was shown with 30 per cent, because rowan trees were often cut down, frequently during the clearing operation. Juniper trees were saved in almost all the cases.

The degree of environmental protection evaluated in this investigation has been far from "considerably damaging present land use" as is the economic limit outlined by the Forestry Act. Conflicts with economic interests are, however, still likely to influence the protective measures. This may be especially true for private, individual forest owners whose clearcuts are small, and single retained trees therefore represent a comparatively great value. As long as people react negatively when trees and bushes are retained, there is also a potential conflict between flora and fauna protection and the aesthetic impression of the clearcut.

Logging debris and deep tracks

Logging transports and soil scarification may also consider the retained vegetation an obstacle. Aesthetical and practical reasons may therefore impede much of the flora and fauna protection.

According to the prescriptions issued by the National Board of Forestry:

> Water springs, watercourses, ditches, shores and river banks should be cleared from logging debris, if necessary, to maintain passage for people and animals and to prevent flooding.

Logging debris should also be removed from:

> Commonly used forest paths, tracks and roads, and objects important to history of culture. [7]

In half the areas "no consideration" was shown, implying logging debris was left to such an extent it was considered an obstacle either to the passage of people and animals or to the flow of water. Where "full consideration" was shown, buffer zones along watercourses and lakes were often retained; or the problem was avoided simply by felling the trees away from the water. It is, of course, much more costly to clear away logging debris after the harvest is completed than when the forest machines are still in the area.

The prescriptions are vague when referring to deep tracks and damage from logging transports:

> On easily damaged ground, logging transports should be avoided or be carried out so that a long-lasting damage will not occur. If deep tracks lead to water floods, runoff should be secured. Special care should be showed in recreational areas, where deep tracks should be restored. [8]

The investigation showed no severe damage from transportation on forest paths or roads. Nevertheless, there were frequent damages on soft ground, such as bogs and fens and wet forest areas. Deep tracks were made on 40 per cent of the clearcuts where such problems were expected. There was a significant impact to the logging of wet forest areas, as 70 per cent of the transport damage took place in such areas. The weather was not extremely wet during the periods of investigation.

The multiple regression explanatory model

The explanatory power of different factors contributing to the outcome of environmental protection described above was analysed through a probit model, using multiple regression. The harvest operation system showed to be the most important factor affecting the degree of environmental protection (See Table 4.1).

Table 4.1
Probit coefficients for the degree of environmental consideration
(The independent variables are compared to a manual felling and processing by the Swedish Forest Service in Västerbotten county)

Independent variables	Dependent variables					
	Total degree of consideration	Environmental protection also technically favoured	Aesthetic values	Nesting trees for birds	Other floristic and faunistic values	Logging residues and deep tracks
Mechanized felling and processing	-0.46(*) } -0.42(*)	-0.79(*)	-0.40	-0.92**	-0.12	-1.14***
Manual felling, mechanized processing	-0.39 }	-0.27	-0.56	-0.64*	-0.16	-1.32***
Private company felling on company owned forest	-0.29	-0.12	-0.28	-0.41	-0.26	-0.49
Private felling on private owned forest	0.16	0.30	0.15	1.34*	0.02	0.40
Private company felling on private owned forest	-0.23	-0.49	0.44	-0.86(*)	-0.05	-0.05
Forest Owners' Association felling on private forest	-0.27	1.01	-0.26	-0.02	-0.27	-0.70
Västernorrland county	0.07	-0.35	0.08	-0.48	0.22	0.20
Värmland county	-0.10	0.04	0.65	-0.22	0.04	-0.22
Östergötland county	0.23	0.66	-0.14	-0.07	0.49(*)	0.19
Norrbotten county	0.68	0.33	-0.13	0.13	0.97*	-0.25
Size of clearcut	0.003	0.01	0.01	0.01	0.001	-0.0006

Note: (*) Significant at the 10% probability level (double-sided test)
* Significant at the 5% probability level (double-sided test)
** Significant at the 1% probability level (double-sided test)
*** Significant at the 0.1% probability level (double-sided test)

In general, the explanatory power of the different independent variables was low because of the limited sample. When each factor was treated in a simple bivariate regression analysis, the probit coefficient was generally strengthened. The mechanized harvest operation systems would then be significantly more negative to the degree of environmental protection (at the 5 per cent probability level) compared to the manual. Private felling on private land would, however, no longer be negative compared to the fellings of the Forest Service. The above differences of results are, however, not significant.

The size of the clearcut area did not affect the result of environmental protection when analysed independently of other factors. Even though the amount of environmental protection required before cutting increased with larger size of clearcut (r=0.31), there was no significant difference between large and small clearcuts when the felling had been carried out, according to the regression analysis. The size of the clearcut might, however, affect nature conservation values in other ways, as will be elaborated in the discussion.

Harvest operation system

There was a significant difference in environmental protection achieved depending on the harvest technology (See Figure 4.3). Harvest operations were classified into three categories. "Manual" meant felling and processing with chain saws, while mechanized systems were split in two categories: felling and processing performed by a harvester, and manually felled but mechanically processed. When manual felling and processing were used the total share of "no consideration" was the lowest. The results did not differ between the two types of mechanized harvestings.

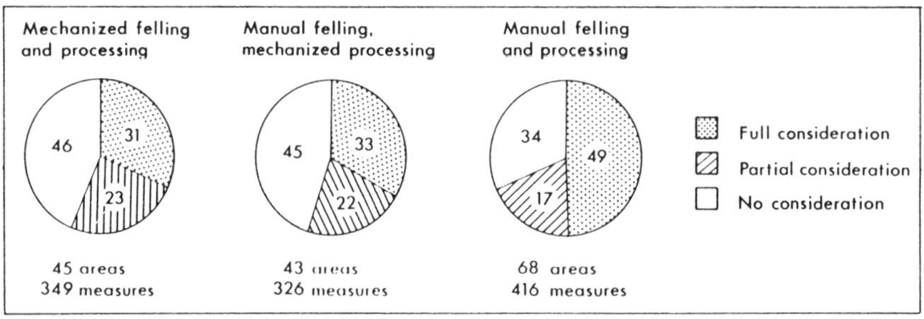

Figure 4.3 Degree of environmental protection by harvest operation system (Per cent of the total number of environmental measures required)

Mechanized harvesting affected all aspects of environmental protection negatively compared to the manual system (See Figure 4.2). Logging residues were more seldom cleared away from watercourses or forest paths, the reason being the large piles of logging residues with mechanized processing; and the forest machines caused deep tracks and damaged the ground. However, these piles could have been removed by the forwarder. The more number of forest machines working on site and the heavier the loads of wood, the more damage was caused to the ground. Most deep tracks were created when the logging transports crossed soft ground and when few haul roads were used. There was also a significant difference between harvest operation systems when saving nesting trees. One explanation may be that nesting trees are frequently "in the way" of the forest machine; it is also possible the nesting trees are more difficult to discover by the machine operator.

Forest ownership and who fulfilled the harvesting

The management policy of the forest owner is assumed to affect the extent of environmental protection considered on his/her forest land; in the same way different forest organisations or individuals who carry out the harvest operation are assumed to have an impact on the outcome of environmental measures protected. Forest owners were categorized into state and private ownership, including both forest companies and private individuals. In addition, a total of four clearcuts were investigated on public land, owned by municipalities and common forests [9]. When forest land was owned by the companies or the Forest Service, the cutting was almost exclusively carried out by the forest owner staff. Private persons as well as the public organisations who own forest land often hired contractors. Hence, it was also interesting to examine those who actually cut the trees as they were the ones directly responsible for the ultimate result of environmental protection.

The multiple regression analysis revealed no general differences between different categories of forest owners and harvest operating organisations (See Table 4.1). The only significant difference was comparing the protection of nesting trees for birds. The private forest owners and managers harvested a larger share of nesting trees than the Forest Service. The private companies also tended to fell more nesting trees on privately owned forest land. Possibly, this is due to economic considerations.

To a small-forest owner, the economic value of nesting trees may be perceived to be relatively high compared to the total stumpage value. [10] When the clearcut is large, as is common for the private companies' and the state-owned forests, the economic value of nesting trees is small compared to the total stumpage value, and may therefore be perceived to be neglectable. In addition, the fixed costs for the harvest (building roads, transporting equipment and personnel etc.) are distributed on a larger number of trees when the clearcut is large. Retained nesting trees on a small-size clearcut may therefore lead to a significant increase in average cost of harvesting the other (non-protected) trees.

Within each forest owner category there was great variation in environmental protection. However, when the total amount of environmental protection was summed up for each category, some differences appear (See Figure 4.4). In total figures the Forest Service protected the environment best, followed by the private forest owners. The private companies showed the least regard to environmental protection. There was no indication of differences between the six forest companies investigated.

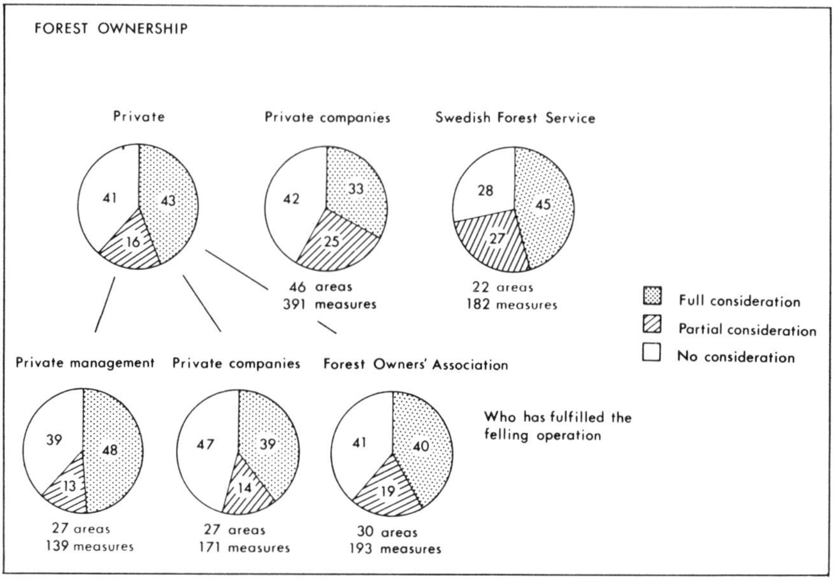

Figure 4.4 Degree of environmental protection by forest owner category and who has fulfilled the harvesting (Per cent of the total number of environmental measures required)

Analysis of the different harvest operators on privately owned land confirmed the lack of sensivity to environmental protection by forest companies. The forest companies used mechanized systems more often than the other forest owners (See Table 4.2). The majority of the investigated Forest Service' clearfellings were manually harvested. Hence, the forest owners' performance of environmental protection was mainly determined by their choice of harvest technology.

Table 4.2
Harvest operation system by forest owner category (Percentage)

Forest owner category	Mechanized felling and processing	Manual felling (chain saws) Mechanized processing	Manual felling and processing (chain saws)
Private	22	24	53
Private companies	46	33	22
Swedish Forest Service	27	32	41
Other public forest	25	0	75
ALL CATEGORIES	30	27	43

There is another important difference between forest owner categories, namely the size of the clearcut area (See Table 4.3). The average clearcut on private, industrial forest land exceeded both the Forest Service and the non-industrial private owners. As will be elaborated in the discussion in the end of this chapter, the low degree of environmental protection at private, industrial clearcuts is especially serious to nature preservation since it affects such a large area.

Table 4.3
Mean size of clearcut (in hectares) by county and forest owner category

Forest owner category	Norr-botten	Väster-botten	Väster-norrland	Värmland	Öster-götland
Private	7.9	2.9	7.1	4.1	4.2
Private companies	101.3	33.0	49.3	10.7	12.5
Swedish Forest Service	71.0	26.5	44.1	-	15.2
Other public forest	-	-	-	-	1.7
ALL CATEGORIES	39.2	15.5	31.5	6.8	6.6

County

The conditions for environmental protection were different in the five counties investigated. Even if economic forestry is practiced in all the counties, the pressure from recreation and nature conservation interests varies. The counties of Östergötland (a relatively highly populated area of Sweden) and Västernorrland stand out as being areas of relative importance to environmental protection (See Table 4.4). Although the average size of clearcuts was smallest in the county of Östergötland, there was evidently a greater need for special environmental considerations. In the county of Norrbotten (the most sparsely inhabited), the number of environmental protection opportunities was the lowest, despite the large size of clearcuts (Compare Table 4.3).

Table 4.4
Frequency of environmental protection required by county

County	Percentage of environmentally valued areas (e.g. of recreational interest)	Mean number of environmental protection measures identified per clearcut
Norrbotten	0	6.0
Västerbotten	11	7.7
Västernorrland	22	9.7
Värmland	8	7.8
Östergötland	43	9.9
ALL COUNTIES	22	8.7

With multiple regression analysis the location of clearcuts in different counties did not significantly affect the result of environmental protection (See Table 4.1), except for the protection of flora and fauna habitats, which tended to be greater in the counties of Östergötland and Norrbotten compared to Västerbotten, Västernorrland and Värmland counties. This indicates a difference in forest-environmental culture between the counties in terms of care for flora and fauna, which are the environmental aspects that probably require the most ecological knowledge to protect. The other aspects, such as aesthetic values, nesting trees and deep tracks, may be more apparent even to a non-educated person. Possibly, the somewhat greater environmental protection achieved in the county of Norrbotten indicates a change of attitudes. The awareness of flora and fauna protection may have increased during the last years due to information and education campaigns. The investigation in Norrbotten was carried out in 1984-1985, whereas the other counties were investigated in 1982-1984.

The frequency of different harvest operation systems varies between counties, partly as a result of different forest ownerships. In the counties of Östergötland and Norrbotten, the manual method was commonly used (See Table 4.5). Hence, the difference in environmental protection could mainly relate to the degree of mechanization. The more manual fellings, the more environmental protection was considered.

Table 4.5
Harvest operation system by county (Percentage)

County	Mechanized felling and processing	Manual felling (chain saw) Mechanized processing	Manual felling and processing (chain saw)
Norrbotten	38	8	54
Västerbotten	24	29	47
Västernorrland	41	34	25
Värmland	50	33	17
Östergötland	16	23	61
ALL COUNTIES	30	27	43

Discussion

Compliance to the Forestry Act was evaluated in two ways. First, in the analysis made so far, the total degree of environmental protection was examined with regard to three aspects of considerations, such as technically/economically favoured, aesthetic, and floristic/faunistic values. The total amount of different environmental features actually protected were compared to the amount of potential protection.

The second way of evaluation was to examine the total share of clearcuttings which conform to the regulation. It was, however, difficult to judge whether a cutting followed the recommendations, since the criteria were vague and an exact limit of compliance to the regulation did not exist. Therefore, the evaluation was made separately for each clearcut area, taking into account the relative ecological importance of each protective measure. The areas investigated [11] were re-evaluated and classified into four groups: (1) failure; (2) failure with some hesitation; (3) passed with some hesitation; and (4) passed. One out of five passed the Forestry Act requirements without hesitation. In addition, two more passed with some hesitation. Two out of five failed to meet the requirements. None of the failures "broke the law", since the Forest Agency had not reacted, nor had it issued any prescriptions for the clearcuts investigated.

The aesthetic conscience of those who work in the woods seems to be significant. Forest areas which improve the aesthetical impression of the clearcut are frequently retained. Perhaps aesthetic reasons also explain the few number of retained single trees, such as nesting trees and snags, since single trees may appear visually obtrusive. If buffers along bogs and fens, watercourses and lakes are not cleared, and tall stumps, dead trees and broadleaved trees and bushes are not felled, the clearcut may not be regarded as "clean". The argument above is commonly heard in forest management discussions. The public reaction to retained snags, windthrows and logging debris is also negative according to attitude studies (Hultman 1983).

During the 1960s and 1970s when mechanization increased enormously in Swedish forestry, management recommendations became oriented towards large-scale harvesting and discouraged retention of vegetation on the clearcut area. A change of recommendations was initiated in the end of the 1970s; the new management strategies emphasize more small-scale biological considerations. Selection of different tree species and different harvest technology according to local variation at the same clearcutting are examples of such small-scale adjustments.

Mechanization, as it is now designed, apparently impedes environmental protection. The comparison made in the multiple regression analysis does not, however, take into account that mechanized clearcuts are much larger in size than manual clearcuts. Even if there were many bogs or many old broadleaved trees on the proposed clearcut, the degree of environmental protection was only registered once for each type of measure. Therefore, comparison with smaller-size areas is somewhat misleading, because the impact of "no consideration" is probably more severe to nature conservation on large clearcuts. Plants and animals, whose habitats are destroyed, generally colonize the area much slower the greater the distance to similar habitats. At small-size clearcuts, colonization thus happens faster because the surrounding forest is near and functions as a refuge.

The clearfellings investigated averaged 27 hectares of fully mechanized cuttings, 18 hectares of partly mechanized, and only 8 hectares of manual cuttings. Nevertheless, it is not fair to simply multiply the degree of environmental protection with the size of the clearcut. If, for example, the same number of nesting trees were present at two different proposed clearfellings and the nesting trees were all cut down, the consequences to potential nestings birds would be the same whether the clearcut area was large or small. Usually, however, there are more environmental aspects to protect when a clearcut is large.

The different results between manual and mechanized cuttings could be explained in two ways. First, either the knowledge or the attitudes to environmental protection may differ between the different groups of workers. Second, the explanation may be purely technical - the machines may not be properly designed to minimize damage to the ground. If the machines were smaller and more flexible, it may be easier to identify and to save single, environmentally valuable, trees and bushes. Such types of harvesters are introduced with thinning operations. [12] The improvement of large baloon tires is an example. In Chapter 5 and 6 the reasons for divergent environmental results with mechanized cuttings are further analysed.

The greater the distance from residential areas the less environmental protection was shown. Near residential areas aesthetic protection was more frequently needed, while in more remote areas flora and fauna protection became more important. However, the remote clearcuttings were more often mechanized. Consequently, the chance of finding large, totally cleaned, clearcuttings increased with the distance from residential areas.

It is difficult to predict how much environmental protection is achieved in other counties than the ones investigated. Since the degree of mechanization varies between counties the degree of environmental protection is expected to vary accordingly. Other factors, such as the "environmental protection tradition" may also be important. For example, the attitudes of different forest owners as well as the attitudes and behaviour of the different branches and sections of national authorities may vary between geographical regions.

Since the clearfellings were investigated, an extensive environmental education programme has been initiated. Not only the national and county agency personnel are educated, but also private foresters and loggers. During the interviews, many claimed a change is taking place implying more concern for flora and fauna protection. The emphasis on aesthetics in forestry practice may, perhaps, be caused by the content of the earlier forest-environmental protection information. In the same way, emphasizing more flora and fauna protection may enhance the sensitivity of forest professionals.

NOTES

[1] This chapter is a summary of my thesis for a licentiate degree in Forest Economics and is earlier presented in Eckerberg (1986).

[2] The model is earlier described in formula (2). Compare Chapter 2.

[3] Many of these clearfellings may, however, be located adjacent to other clearcuts, which may explain why the total clearcut <u>area</u> often exceeds what is described in the statistics. Although the Forestry Act recommends that a new clearfelling should not adjoin an old clear-

cut, as long as the new forest is not "highly visible", this is not always observed. Instead, private forest owners often take the chance of cooperating when cutting their forest. Large-forest owners often continue cutting adjacent to the old clearfelling because of windthrows along the edges. Consequently, the clearcut areas are often much larger than reported on the clearcut notifications.

[4] The extent of the feature was noted on a separate form during the field investigation to facilitate post-harvest evaluation, but in the computer analysis it was registered before harvest as if the feature was present or not present.

[5] "Young stands of trees" refer to small groups of young trees, covering too small an area (smaller than 0.5 hectare) to be treated separately within an old forest stand.

[6] Diametre Breast Height.

[7] NBF regulations 1983:2 to the Forestry Act § 21.

[8] Ibid.

[9] Common forest are explained in Chapter 3.

[10] Stumpage values is the price paid for the standing timber, usually computed as the price paid for logs at a mill minus the cost of transporting timber to the mill.

[11] Not all 158 areas were re-evaluated in this way, since different judges made the primary evaluations. Only the 82 clearfellings that I investigated in the field were re-evaluated (Compare Table 2.2).

[12] The reason why small and flexible machines are used in thinnings is that if large machines are used, the remaining trees are often severely damaged. The roots are cut off by the deep tracks, and in spruce stands (Picea abies) the noxious root-rot (Fomes annoses) frequently spreads into the stand. Consequently, the thinning operation may contribute to jeopardizing the remaining trees instead of promoting their volume and quality growth.

5 Case studies of failure and success

In some cases most environmental features at a clearcutting were protected; in other cases the clearcut area was cleared of all environmentally valuable trees and bushes, and deep tracks and torn ground vegetation were prevalent. In this chapter, the relationships between dependent and independent variables are described in connection to specific case studies of clearfellings. The aim is to give a picture of what happens in reality, and to illustrate how difficult it may be to establish a few necessary conditions which are to be fulfilled for predicting policy outcome.

Starting from the case studies of clearfellings, the people involved in the harvest operation were identified and interviewed about implementing environmental protection. With this bottom-up approach, it is assumed that the "street-level bureaucrats" at the lower level of hierarchy exercise substantial discretion in their work and that their behaviour, to a large extent, actually determines policy outcome (Lipsky 1980). Moreover, implementation is believed to be very much affected by the way in which the target groups react to the policy, that is the people whose behaviour constitutes the "bottom" of the implementation process. Their decisions cannot be standardized, managed, or controlled in all situations using conventional administrative tools. Varying circumstances in different cases requires judgement, skill and imagination (Elmore 1979). Hence, it is important to examine what is the response of the forest professionals to the environmental requirements, and what other factors in their environment may affect the way in which they protect - or do not protect - environmental features.

Characteristics of the total twenty-five clearcuts which were followed-up by network interviews are summarized in Table 5.1. The cases analysed in detail in this chapter were selected from the 158 clearcuttings which had previously been investigated in the field. Since they are not a random sample, they are not representative for Swedish forestry in the sense that they can be used to show the share of different forest categories or different harvest technologies. Nonetheless, the varying characteristics of the twenty-five clearcuts illustrate the natural divergence in environmental protection conditions. In addition to merely illustrating the implementation process from a bottom-up point of view, the case studies suggest multiple factors which, to various extent, could explain the result of environmental protection. The explanatory factors were presented by the actors at the lowest level of implementation: the ones who planned the clearcut operation, supervised it, and the ones who actually cut and transported the trees.

In each of the case studies, there was one person who, to a larger extent than the others, determined the outcome of environmental protection. He could be characterized as the "key actor", since his decisions substantially affected the behaviour of the other actors. Therefore, his attitudes to environmental protection, his forest economic considerations and his contacts with environmentalists may be important factors that contribute to the harvest performance.

The summary of all twenty-five case studies shows that the characteristics of the environmental protection outcome, of the harvest operation and of the person who mostly determined the result - the key actor - varied substantially.

Three case studies were selected from this material, in order to further illustrate how different circumstances contribute to the outcome of environmental protection. Criteria for selection were that the cases represented different degrees of environmental protection achieved, ranging from "failure" to "success". These three case studies are presented in detail.

The first describes a cutting where very little regard was payed to floristic and faunistic values, and the total result of environmental protection was very poor (=case # 23 in Table 5.1); the second case shows a clearfelling where aesthetic values were neglected but flora and fauna features adequately retained (=case # 13), and the third a case where all important aspects of environmental protection were considered in accordance with the regulation (=case # 9). After the three cases are presented, the results from all case studies are summarized and discussed.

Table 5.1
Characteristics of twenty-five case studies of clearcuttings

Case #	Environmental protection outcome		Characteristics of the harvest operation			Characteristics of the key actor			
	Total degree of environmental protection[a]	Protection of flora and fauna[a]	Number of people involved[b]	Forestry Agency influence	Harvest technology[c]	Duty of work	Economic considerations[d]	Attitude to environmental protection[d]	Contacts with environmentalists[e]
1	average	low	7	no	mechanized	Timber officer/planner (Forest Owners' Ass.)	large	low	moderate
2	very low	very low	6	no	mechanized	Timber officer/planner (Forest Owners' Ass.)	large	low	none
3	average	average	6	no	mechanized	Timber officer (Private company)	-	-	-
4	high	average	8	no	mechanized	Forester (Forest Service)	large	moderate	moderate
5	very high	average	6	no	manual	Forester (Forest Service)	moderate	high	frequent
6	average	low	5	no	mechanized	Forester (Private company)	moderate	moderate	moderate
7	low	very low	3	no	mechanized	Timber officer (Private company)	moderate	moderate	moderate
8	low	very low	6	no	mechanized	Timber officer (Forest Owners' Ass.)	large	low	moderate
9	very high	very high	7	no	mechanized	Forestry society forester/ marking of trees	low	high	frequent
10	very high	high	4	no	manual	Forester (Forest Owners' Ass.)	moderate	moderate	frequent
11	very high	very high	5	yes	mechanized	Forestry Agency foreman/ marking of trees	moderate	moderate	none
12	average	low	8	yes	manual	Forestry Agency foreman/ marking of trees	-	-	-

13	average	high	5	yes	mechanized	Forestry Agency officer/ marking of trees	low	moderate	none
14	average	low	6	yes	mechanized	Forester (Forest Service)	moderate	moderate	moderate
15	high	high	8	no	mechanized	Timber officer (Private company)	moderate	moderate	moderate
16	average	average	7	yes	mechanized	Timber officer (Forest Owners' Ass.)	large	moderate	none
17	average	average	5	no	mechanized	Timber officer (Private company)	moderate	high	moderate
18	low	low	4	no	mechanized	Private forest owner	low	moderate	none
19	very high	high	4	no	mechanized	Timber officer/planner (Private company)	-	-	-
20	very low	very low	5	no	mechanized	Timber officer (Private company)	moderate	moderate	moderate
21	high	average	5	no	mechanized	Forester (Forest Service)	moderate	moderate	none
22	average	low	5	no	mechanized	Forest workers (Private company)	moderate	moderate	none
23	low	low	4	yes	mechanized	Timber officer (Private company)	large	low	none
24	average	low	4	yes	mechanized	Forester/marking of trees (Forest Owners' Ass.)	large	low	none
25	low	average	6	yes	mechanized	Forestry Agency foreman/ marking of trees	large	high	moderate

Note: a) very low = < 20%; low =< 40%; average = 40-60%; high = > 60%; very high = > 70%
b) excluding Forestry Agency personnel
c) "partially mechanized" with manual felling is included in "mechanized"
d) in comparison to other respondents
e) including with EPS personnel/in comparison to other respondents/moderate = at least once

Questions which are elaborated in this chapter include: What were the forest conditions and the state of environmental protection before and after cutting? What criteria determined the design of clearcutting? How was the harvest organised; what and how many persons were involved? Were there any "key actors" who, to a larger extent than others, determined what environmental features should be preserved? What were their attitudes to environmental protection, their forest economic considerations and their contacts with environmentalists? How did they and the other implementing actors explain the result of environmental protection actually achieved?

Case one: low degree of environmental protection

The harvested area was located in a sparsely populated region in the middle of Sweden, dominated by forestry. A large forest company owned the land. Fourteen hectares of mainly Norway spruce along a slope were clearcut (See Figure 5.1).

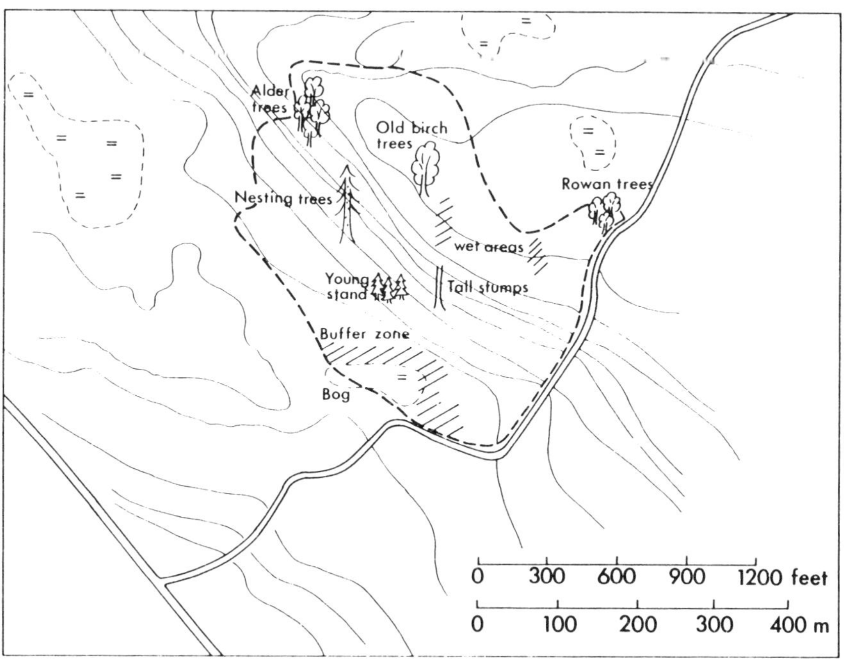

Figure 5.1 Environmental features of case one

Environmental protection features included a couple of wet forest areas in small ravines along the slope, one with small Scots pine at the lowest part of the area. Such non-productive land should be retained according to the environmental regulation, including a buffer zone along the bog with a variation in tree heights and species. Nesting trees, mostly of spruce but also some tall, birch-stumps with holes were detected before cutting commenced. Birch and spruce snags, old birch and alder were scattered on the area. A group of around thirty alder trees grew on the north-western part of the forest slope. Three rowan trees and old windthrows were also detected before harvesting. Broad-leaved trees and snags are environmentally valuable because many rare plant and animal species depend on old wood in different stages of decomposition. Old broad-leaved trees are, for example, important hosts for lichens, mosses, insects and birds.

After the harvest, few of these environmental features were left. The pine bog was partly harvested and no buffer zone was retained. The small ravines of wet forest were cleared. Deep tracks were prevalent in wet areas. Haul roads passed across sensitive areas, thereby creating small pools of water and torn vegetation. Except for one tall spruce snag, all nesting trees were gone. The rowan trees, all birch and alder trees were also harvested. Windthrows were, however, left on the ground - the majority were already decayed and thus not usable as pulp wood. A group of young spruce was retained, offering at least some wildlife protection.

The forest company owns land over a vast area. The forest is geographically divided into districts, where a forester supervises forest management. Forestry planning is carried out at the district office; the yearly management plans are based on demands from the company headquarters on yield from clearcuttings and thinnings, derived from economic goals and forest conditions. Although there are instructions from the headquarters on what areas are appropriate to cut, the detailed planning of forest operations is left to the district planners. Clearcuttings are spread geographically, according to the forester who was interviewed, in order to avoid too large sizes of clearcuts. After the most suitable forest areas are selected from the forest data base, operational planning is carried out by the timber officer. He is usually provided with approximately thirty per cent surplus of proposed clear-fellings for every season, thereby leaving room for additional selection. The timber officer inspects most areas in the field before logging commences; half of these areas are inspected together with the district planner.

Before harvesting commences, the timber officer marks the outer borderline of the area to be clearcut with plastic bands. Usually, if the felling is made with a harvester, small trees and bushes are cleared away by a team of forest workers with spacing saws before harvest begins. Thus, the design of the area to be clearfelled is already established at the time when the harvester arrives. The forest workers who operate the machines only have to comply with the instructions from the timber officer and follow the marked borders. After the trees are felled and processed, a forwarder collects and transports the logs to a landing. Because of high investment costs for the large machines, they are utilized as much as possible; two or three forest workers relieve each other on working shifts. One to three years after the harvest, soil scarification is usually carried out with a scarifier.

The District Forest officer said he had been to the area before it was harvested. He had no objections to the cutting, and did not discuss the planned harvest with the forest company. Hence, he was aware of what the forest was like - he knew it was to be clearfelled, but did not make recommendations.

> We receive around 200 clearcut notifications every year and inspect maybe half of these areas before or after harvest. The company forests are more seldom inspected compared to those owned by private persons. This year we have, however, initiated an objective sampling of regeneration inventories also on company-owned clearcuttings. I could never monitor my district of 100,000 hectares without my profound local knowledge. I was born in this area. (District Forest officer) [1]

This area was selected for clearcutting by the district planner together with the timber officer from the forest company district. The size of the clearcut was moderate compared to the company's earlier, much larger clearcuts:

> The clearcuttings used to be larger at the time when we still practiced controlled burning [2] and throughout the 1960s. When the public started to protest, we changed into smaller-size clearcuts. (Company forester)

> There is no economic advantage of felling areas larger than fifteen hectares; therefore, there is no reason for us to clearcut larger areas. (Forest planner)

The timber officer marked the border of the felling, but made no specific instructions to the forest workers to protect any environmental features. The bog was harvested on purpose in order to drain the area and make it more productive.

> Earlier, we have sometimes retained wet forest areas. As a result, the remaining trees have died from excess water, and the Forestry Agency have requested us to cut them because of the risk for insect damage. (Timber officer)

According to the timber officer, he had earlier told the forest workers in general to try to leave nesting trees and some dead trees when harvesting, for environmental reasons. However, no specific instructions were made in this case. He had not recommended protection of broad-leaved trees:

> Rowan trees are sometimes left... but they might be "in the way" [of the harvester]... We do not retain large birch trees, as in this case... Alder trees are not retained unless they grow on wet land. (Timber officer)

Felling and processing was performed by a harvester, followed by a forwarder which transported the logs to a temporary landing along the road. The forest worker who operated the harvester claimed he rarely had any instructions to protect environmental features:

> The border of the cutting is marked, but the rest should be clearfelled... Trees on bogs and other wet areas are felled to the extent that the harvester can reach them without getting stuck... Nesting trees are left if we detect them... I can't remember having seen any rowan trees on this area... Broad-leaved trees are usually cut, but sometimes we have been requested to leave some aspen trees. Windthrows are utilized as long as the wood is not too decayed... (Forest worker)

The forest worker had just recently begun to study environmental protection in a study circle organised by his labour union. As he remarked, his environmental conscience was mostly influenced by mass-media reports. He did not believe there had been any important environmental values which should have been protected in the forests where he had worked so far. Nesting trees was the only specific measure he mentioned as important to consider when operating the harvester.

The conclusion from this case study is that the timber officer had a key role to the result of environmental protection. He determined the design of the clearcutting, and did not point out any specific environmental features to be protected. His decisions were, however, influenced by the management policy of the forest company. The forester in charge of the district could have specified environmental directives, in the same way as he had advocated a limit to the size of clearcuts. Generally, the company personnel gave environmental protection low priority compared to economic interests in forestry. This attitude was clearly expressed by the forester:

> Principally, those who are affected by public interests [such as environmental protection] should be economically compensated. (Company forester)

The harvest technology used here contributed, to a large extent, to very little care for environmental protection.

If the cutting would have been carried out manually, the forest workers would probably have retained more of those trees that were of little economic value, but took time to cut. Most likely, they would have left much more of the wet areas uncut, which also would have diminished the damage from deep tracks. However, manual harvest technology would not alone have created sufficient conditions for a higher degree of environmental care. The instructions from the supervisors and the attitudes of the forest workers would still have been important factors.

It seemed as if the forest company had, to some extent, recognized the need for more protection of environmental values, since the company sponsored the study literature used in the environmental-protection study circles for their staff. The low priority to environmental interests could, possibly, also change if environmentalists would put pressure on the company to change their behaviour, because as the forester had admitted, pressure from the public had been a major reason for reducing the size of clearcuts. So far, there had been no specific reactions from local environmentalists and others who might disagree with the outcome of environmental protection. The Forestry Agency could also have reacted, but as the attitudes of the District Forest officer coincided with the attitudes of the forest company, no premonitions from the government forest authority could be expected.

Case two: average degree of environmental protection

The private forest owner in this case could be regarded as an average forest owner of the region. He owned a total of 250 hectares of forest land in the south of Sweden, located in a fairly populated area. Adjacent to the two hectares which were clearcut, there was a summer cottage, formerly a small farm (See Figure 5.2). The old barn bordered the forest. The summer cottage was rented to an old man and his family; it was owned by the same forest owner. The forest owner lived on his holdings, but worked in town as a public officer.

Dominant tree species were Norway spruce, with around ten per cent of the stock consisting of Scots pine. Two small young stands of forest adjoined the proposed clearfelling. These young stands were excluded from the cutting on the map which was drawn on the clearcut notification. According to the environmental recommendations, a buffer zone towards the buildings should also have been retained, especially a group of pine and birch trees adjacent to the old barn. Other environmental features consisted of two tall stumps (birch and spruce), two spruce snags, a group of hazel and white-beam, rowan and young oak.

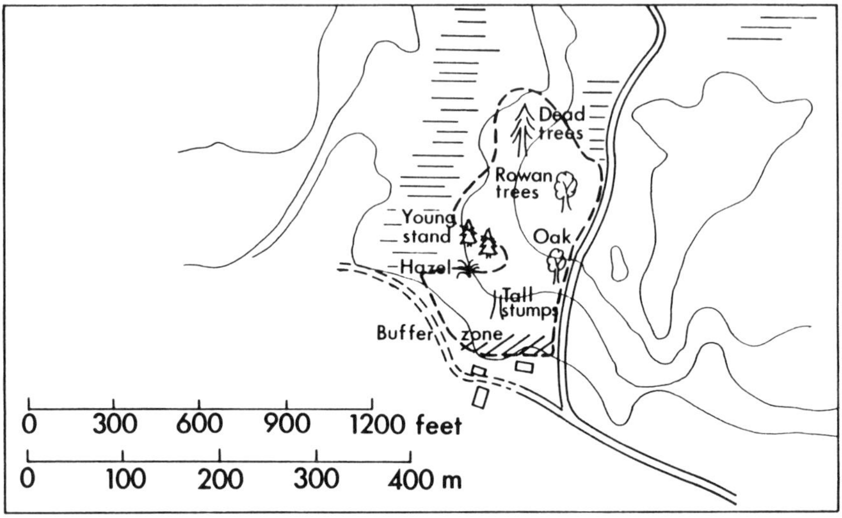

Figure 5.2 Environmental features of case two

After clearcutting, no buffer was protected towards the summer cottage. Only a couple of minor birch trees were left, and some large broad-leaved trees which grew almost in the garden. On the rest of the clearcut area, most broad-leaved trees were retained, including hazel, white-beam, rowan and oak. The snags and tall stumps were harvested. No deep tracks were detected and the forest path along the eastern side of the clearcut was neat and clean.

The forest owner was aided by a private forest consultant. The consultant and the forest owner had walked through the area and discussed what operations were suitable. At this occasion, they decided to clearcut the area, and they marked the border of the cutting. Thereafter, the forest owner notified the County Forestry Agency and hired them to estimate the total timber yield from the proposed clear-felling. A Forestry Agency officer inspected the area, marked all trees which should be cut, and presented the volume figures. The harvest yield was then sold to a private saw-mill, who in turn hired a contractor to carry out the harvest. Felling and processing was performed by a harvester. A forest worker from the saw-mill transported the logs with a forwarder to the landing located next to the barn.

Besides the Forestry Agency officer who marked the trees, the District Forest officer had inspected the forest before clearcutting commenced. The DFO said he inspected approximately seventy per cent of all notified clearfellings in the field; however, the emphasis on these inspections was not on environmental measures but on examination of the age of the forest, to assure that immature forests were not clearcut.

Sometimes he had given specific advice to forest owners about environmental protection, but not this time. The Forestry Agency officer who marked the trees had not discussed environmental aspects with the forest owner, nor had he considered retaining a buffer zone near the summer cottage.

> Adjacent to the summer dwelling we only marked economically valuable trees [for cutting]. The rest was left ... We never discussed retaining more trees. (Forestry Agency officer)

When the DFO saw the pictures from the clearcutting during the interview, he was quite upset over the lack of buffer. To his opinion, they should have retained at least a group of trees close to the cottage. The forest owner did not agree; he said the inhabitants of the cottage had no objections to the buffer being cut down.

> In that case there should have been more broad-leaved trees [from the beginning], but now it was mainly spruce. Some people like to have the forest in their close vicinity. Others prefer the area totally cleaned and the trees cut down it all varies ... (Forest owner)

The forest worker's instructions were to harvest all marked trees. Therefore, he also felled a big birch tree behind the barn, which he remembered very well - he nearly damaged the building. The forest worker's opinion was that such trees should never be harvested, but left as a shelter. He said it was generally impossible to leave any trees that had been marked for felling. The saw-mill had already paid for them and the forest owner would think he did a bad job if the area was not clean. Small trees and bushes were not marked; therefore he did not cut them. Consequently, most broad-leaved trees were left on the clearcut area. The snags had been marked by the Forestry Agency officer and were thus harvested; the tall stumps had not been marked, but according to the forest worker they were not deliberately cut. Perhaps, the tall stumps had been overthrown by surrounding trees when those were harvested.

The design of this clearcut was mainly planned by the forest owner and his advisor; however, in terms of environmental protection, the Forestry Agency officer was largely responsible. The officer should have been aware of the environmental protection requirements of the Forestry Act, and could have influenced the protection of the buffer zone and the snags. The saw-mill company and the contractor depended on the will of the forest owner. The saw-mill company could hardly be held responsible for the result, other than if the harvest operation had been (which is was not) carried out in a careless way. Except for the DFO and the forest worker, who were critical of cutting of the buffer zone, all the actors involved were unanimous about the appropriateness of the clearcut. The old man who rented the summer cottage was not interviewed of practical reasons. Most summer visitors would probably be upset about the clearcutting, at least during the first years after harvest, since the clearcut area expanded all the way to the garden.

The harvest technology did not, to any large extent, jeopardize the environment. A manual forest worker would probably also have cut down the buffer zone, where the trees were marked for felling. Instead, this case highlights the importance of positive attitudes toward environmental protection among all those involved. Forest economic considerations probably also contributed to the forest owner's will to harvest the buffer strip. The relatively high degree of flora and fauna environmental protection in this case, was achieved without any outspoken intention to protect such values; rather, the broad-leaved trees and the tall stumps were left because they had no economic value.

Case three: high degree of environmental protection

The municipal-owned forest area to be clearcut was located in an area with great environmental interests. Through the forest runs a small stream, which according to the Nature Conservation Plan is one of the county's two most important spawning areas for salmon-trout. In addition, the fault where the area is located is pointed out in the national inventories as an area of great geological interest. The Nature Conservation Plan states that

> the environmental values will remain if the stream is not affected by, for example, excavations, buildings, water pollution or use of water for irrigation ... [and that] forest cuttings should be prevented in the steep slopes and on the ridges.

An area of 2.3 hectares were to be clearcut, framed by a small road on the west side, leading down to a village on the coast, and the stream on the east side (See Figure 5.3). The forest consisted of mostly old Norway spruce, but also scattered Scots pine and broad-leaved trees. Alder trees were concentrated along the stream and on the slopes of the small ravine, together with hazel. Birch, rowan, juniper and rose-hip bushes created an inter-mediate zone between the tall spruce forest and a pasture. Three houses were affected by the proposed cutting: one summer cottage next to the pasture with its garden adjacent to the forest; a second small summer house adjacent to the forest on the highest point - and thus with a good view over the forest; and a larger house, also to some degree overlooking the forest but from a greater distance, which was used in the summertime as a children's holiday camp.

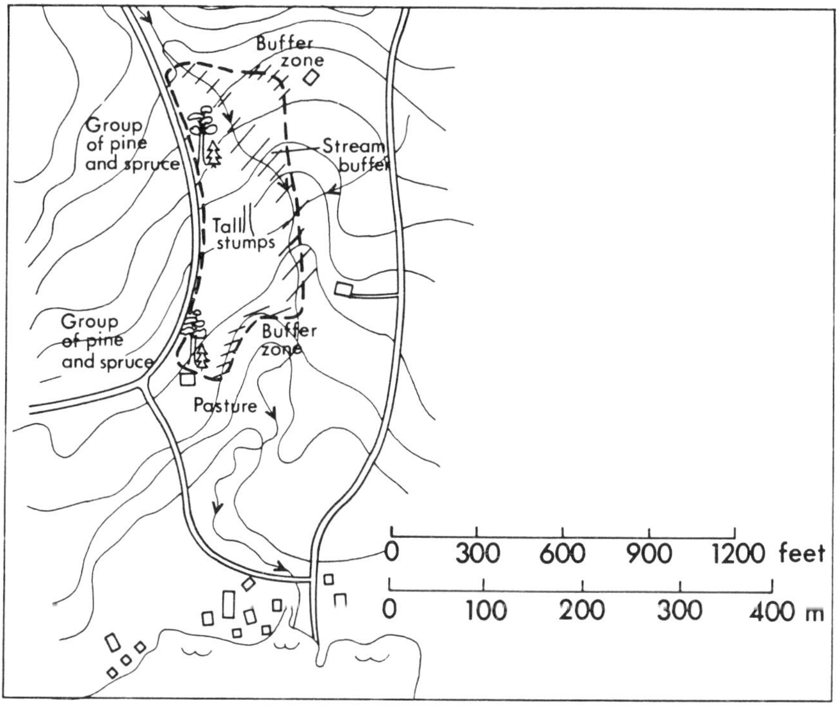

Figure 5.3 Environmental features of case three

The most important environmental features to protect at this cutting was the stream buffer zone, being of great value to flora and fauna. Aesthetic environmental protection was also in question. Old spruce are usually difficult to retain in buffer zones because they often blow down. At this cutting there were, however, enough pine and broad-leaved trees and bushes to create suitable buffer zones. If generous groups of such trees were retained, the clear-cutting would comply with the environmental requirements. In addition to the environmental measures already mentioned, there were a few other specific features to protect: one nesting tree (an old birch with holes), a tall birch stump, and approximately a dozen of conifer snags and old wind-throws.

The majority of suggested environmental features were protected with clearcutting. The stream was well protected, with the alder trees and hazel retained, including spruce in the ravine. Aesthetic groups of pine and spruce were left; one large group along the road and another adjacent to the summer cottage besides the pasture. No deep tracks or damage to the ground vegetation were detected. Nevertheless, some features were less protected. No buffer was retained towards the upper summer house and logging debris was left on the lawn; the buffer along the pasture was partially cleared; and most of the snags, windthrows and nesting trees were cut. Despite these "small protection measures" not considered, the total degree of environmental protection was very high compared to other clearcuttings. Because of the large buffer along the stream and the groups of trees retained, the overall impression of the clearcutting was not obtrusive and the view from the summer camp house was not seriously altered.

Although the municipality owns the forest, management of the forest is carried out by the Forestry Society. They administrate the forest holdings, plan, and monitor forest operations. Some clearcut operations are, however, contracted to other forest organisations. The trees to be cut are marked in advance and sold on the stump. Usually, if cutting will be performed by a harvester, pre-harvest clearing of small trees and bushes is carried out by the Forestry Society, which happened in this particular case. The stumpage sale was thereafter sold to a private company, who carried out the operation with a harvester. The transport of logs within the clearcut area was performed by a forwarder from the Forestry Society. Afterwards the wood was handled by the forest company. Deep tracks were avoided on the clearcut because the harvest was carried out during the winter, and also because the ravine slopes were never cut.

The forester from the Forestry Society was fully aware of the sensitive location of the proposed clearcut; therefore, he contacted the county Environmental Protection Section. The forester and the EPS officer inspected the area together, discussed the boundaries of the cutting as well as the extent of stream buffer, and marked all boundaries with plastic bands. The regional inspector of fisheries was also consulted, but had no objections to retaining the stream buffer recommended by the EPS officer and the Forestry Society forester in order to protect the fish habitat. The design of the stream buffer was thus decided by the Forestry Society forester in consultation with the EPS. Remaining environmental measures were decided by the forester alone. The District Forest officer did not inspect the area before harvest, although he would have liked to. He knew, however, that the EPS had been consulted in this case:

> We used to be two [DFOs] on the district. Then it [to inspect forest areas in the field] went pretty well, but now I don't have enough time. I also have carried out many markings this year, so I haven't had any time left over. I wish I could do more... I receive around 200-300 clearcut notifications [each year]. The longer you stay at a district, the more you know. But 30-40 per cent you want to inspect in advance. Then you return [after the harvest] to check regeneration. (District Forest officer)

A team of forest workers from the Forestry Society cleared the area before harvesting, instructed by the forester to make the buffer strip towards the dwellings and the pasture "look nice". The buffer was thus retained for aesthetic reasons and not for the flora or fauna (which explains why quite a few small trees and bushes were cleared away). Much of the environmental result of the cutting was in accordance with the forester's view. He said windthrows are utilized as long as they are good pulp wood, and the rest of them often cut in order to make the way for the forest machines. Snags are retained to some extent according to the forester. Generally, the forester's attitude was quite sensitive to environmental protection. For example, he claimed around one hectare of environmentally valuable land should be possible to retain without economic compensation, that clearcut sizes should not exceed ten hectares, and that in environmentally valuable areas he would usually cut manually in order to protect the environment. He valued both aesthetic, recreational and faunistic aspects, although the aesthetic ones were emphasized. For example, he described the retention of trees near the dwellings as follows:

> We retained broad-leaved trees in order to make it look nice. But the spruce are left by mistake - it doesn't look good now. However, the pine trees were supposed to be left from the beginning. There was nothing to save at the upper cottage - it looked too untidy to protect. Sometimes we discuss [the harvest] with the owners of the cottages, but we usually don't make great efforts to look them up. (Forestry Society forester)

The District Forest officer was critical of the lack of vegetation buffering the upper summer cottage: [3]

> They should have retained more trees near the upper cottage. If there was mostly aspen, it is not very nice, but the birch trees should have been left. They [the birch trees] don't restrict forestry. (District Forest officer)

When the harvester arrived, operated by two forest workers on shifts, all trees for cutting were already marked. The forest workers were instructed to cut all marked trees:

> Yes, the trees there [towards the upper cottage] were also marked - probably, the aspen could be utilized for pulp. Every single bush that could possibly turn into something was marked. Often they also mark trees that will not be usable [for pulp wood]. (Forest worker)

The only environmental instruction communicated to the forest workers through their timber officer was to leave the stream buffer untouched. As the forest workers told, the stream buffer originally was planned to be broader than actually happened. Parts of it blew down while the harvester was operating on the area. The timber officer from the forest company thought it would have been best to protect the entire area, but this opinion was dependent on who was the owner of the forest:

> If our forest company had owned the forest, we would have cut it all. But now, when the municipality owns it they shouldn't have cut at all, I think. But then I disregard the economics. In that case [if you consider economics], they should only have thinned the area. Now they opened up for the wind. Especially the vicinity to the children's vacation camp should have contributed to the forest being protected.

The forester from the Forestry Society had a key role in the design of the clearcut, having advised with environmental authority. Nevertheless, there was critique coming both from the Forestry Agency and the forest company personnel on the design of the buffer towards the upper cottage. In addition, the timber officer from the forest company suggested the entire area should have been protected.

This case highlights a general problem in environmental protection. Many opinions can exist on what environmental protection really is and to what extent it should be provided without economic compensation. The differences in opinions originate from different aspects of environmental protection being valued; for example aesthetic, social/recreational or "pure" floristic/faunistic aspects. It also depends on how the economy of the forest is valued - as an asset for the forest owner alone or as a social resource. In this case the economic value of the forest was calculated by the value of the timber. If, however, the forest value was also calculated by the appreciation of affected neighbours, such as the summer cottage inhabitants and the children on vacation, clearcutting may not have been appropriate at all. Although their municipality owned the forest - and maybe could be expected to pay more regard to the views of its citizens than would other forest owners - these people were never heard. As the forest company representative made clear, even less protection would probably have been achieved if they had owned it. Favourable attitudes towards environmental protection and consultation with the environmental authorities contributed to the comparatively high degree of environmental protection achieved. Still, the case shows that forest economic considerations impeded such protection.

The harvest technology, in this case, did not really matter. Maybe one or two more snags could have been left with a manual harvesting. Otherwise, manual forest workers would probably also have cut those trees that were already marked and retained the others.

Discussion

No single factor seems to determine the result of environmental protection. On the contrary, different specific features characterize the different cases of clearcutting. In the following, some conclusions are made from the three case studies presented above, as well as from the total case material.

Most of the case studies analysed with interviews were performed with harvesters, or with manual fellers and mechanized processors. All three clearcutting cases described in detail were performed with a harvester. They show that mechanized harvesting could be used successfully, with regard to environmental aspects, if only the environmental measures are well planned for in advance. Generally, however, larger machinery probably affects the result of environmental protection negatively, but to various extents in the different cases. [4]

In all the cases, a large number of actors were involved in the implementation at field level. Although the number of actors varies between the different cases, there is no indication that the environmental protection outcome depends on the number of actors involved in harvesting. However, a large number of implementing actors requires co-ordination and clear directives in order to maintain environmental protection measures all the way from the government authorities, or from the forest-owner supervisor to the forest workers. If all actors are not in agreement with the measures, or if they are ignorant, successful protection of environmental features is less likely to be achieved, whereas a single forest owner who carries out his own forest operations will follow his own decisions.

The analysis shows that different categories of workers play the key role in determining what specific environmental features should, or should not, be protected in the different case studies. The Forestry Agency officer was much involved in some of the cases - he actually played a key role to the outcome of environmental protection in four of the twenty-five cases, including "case two", which was described in detail. In the majority of cases, the Forestry Agency did not directly influence the environmental aspects of the cutting. Many of the District Forest officers expressed they had no time to inspect the area. [5]

The majority of key actors were foresters or timber officers who supervised and planned the harvest operation. Some of the forest companies used their forest planners to design the details of eventual environmental considerations. Their decisions then determined, to a large extent, how the subsequent timber officer and forest workers behaved. In quite a few cases, the forest area to be harvested was sold on the stump and all trees marked in advance. Here, the person who marked the trees had a great impact on the amount of trees protected for environmental purposes. In only one case were the forest workers mostly responsible for the ultimate result. Their forest company generally delegates the detailed planning to them, instead of having a timber officer mark out the details of the clearcutting. There was also only one case where a forest owner could be regarded to be the key actor. The forest owners may also, of course, affect the attitudes of their personnel. For example, the forest management objectives of the different forest owners set the limits for how much environmental protection could be carried out in practice, according to several of the foresters and timber officers interviewed.

The key actors' attitudes to environmental protection, their forest economic considerations and their contacts with environmentalists and environmental authorities were specifically examined. (In three of the twenty-five cases, the key actors were never interviewed, and their opinions could not be evaluated.) Low priority to environmental protection and large forest economic considerations of the key actors seems to coincide with low, or average (which in reality implies only around half of the required environmental features were protected) degree of environmental protection. Opposite circumstances, that is favourable attitudes to environmental protection along with low priority to forest economic considerations do not, however, automatically lead to a high degree of environmental protection, according to the analysis of case studies. Apparently, factors other than the key actors' attitudes may work together, or counteract, during the implementation process.

In general, few of the key actors had any contacts with environmentalists. "Moderate" contacts in Table 5.1 therefore implies only one or two such contacts ever. Consequently, the influence from environmentalists and from the environmental authorities (which are included in the list of environmental contacts in the table) cannot be substantial. However, the cases of clearcuttings analysed were neither selected from regions or areas with especially great potential conflicts between economic forestry and environmental protection, nor were they selected from environmentally valuable forest areas as marked in Nature Conservation Plans. The direct influence from environmental agency personnel as well as from the public and local environmentalists is, probably, greater in such areas.

In three cases where the key actors frequently had contact with environmentalists and environmental authority personnel, the degree of environmental protection was very high. This situation could be explained in at least two ways. First, the forester may initiate - or encourage - contacts, and the positive correlation is then mainly explained by the forester's own attitudes. Second, pressure from environmentalists and environmental authorities may also affect the behaviour of the forester to achieve more environmental protection. Probably, both of these explanations are true. Frequent contacts between forest professionals and environmentalists most likely contribute to a higher degree of environmental protection.

NOTES

[1] All quotations are verbation (freely translated from Swedish) with the exception of false starts (oh well.., um...).

[2] Controlled burning was used with the same aim as today's scarification - the ground vegetation was partially burned away, which left the new seedlings with less competition, more nutrients available and a favoured microclimate. This controlled burning has been abandoned by Swedish forestry today - mostly because it requires a lot of personnel who can control the fire.

[3] As previously mentioned in Chapter 2, I showed photographs from the clearcuts during the interviews. The DFO could therefore react to the environmental result without having been to the area himself.

[4] As earlier examined, retention of single trees (such as nesting trees) is often more difficult to achieve with large forest machines (See Chapter 4). The influence of harvest technology is further discussed in Chapter 6.

[5] The role of the Forestry Agency is further analyzed in Chapter 7.

6 Implementing actors' attitudes to environmental protection

More than one hundred interviews were carried out with both Forestry Agency officers, forest owners, foresters, timber officers and forest workers and all had valuable information to contribute to the analysis of environmental outcome. Still, it is difficult to establish measures on the explanatory power of the different independent variables delineated in the introduction (See Figure 1.2). First, the dependent variable - the degree of environmental protection - contains many different aspects in itself; the environment is characterized by ecological relationships and multiple environmental features, of which some were protected and others not. Second, the independent variables are interrelated in a complex network of relevant circumstances. These circumstances are difficult to isolate into a few factors. This is a common problem in social science; causal relationships are generally hard to measure because many factors together, in a variety of different constellations, make up necessary (and sometimes sufficient) conditions for changes according to a certain policy (Hermerén 1984).

The causal relationships between explanatory factors and policy outcome are examined through multiple sources, which is a methodological advantage (See Chapter 2). The implementation of environmental protection is already analysed through field investigation of clearfellings (See Chapter 4). Chapter 5 described the case studies and analysed the multitude of different circumstances and conditions for protecting environmental features with each of these cases. In this chapter, the actors' perceptions of different factors which may affect environmental protection performance are discussed at a general level: (1) implementing actors' attitudes to and knowledge of environmental protection; including the role of education; (2) pressure from environmentalists; (3) the effects of different harvest technologies; and (4) forest economic considerations. The results presented in this chapter are based on the network interviews with implementing actors. [1]

One explanation to the result of environmental protection achieved could be that the attitudes towards environmental protection differ among the different implementing actors, thereby affecting their forest management behaviour. If the people in charge of translating the environmental requirements into forestry practice are to care for the environment, they are likely to do so more frequently if they are well informed, motivated and positive towards environmental protection. The hypothesis above generated a series of questions which are elaborated in this chapter. The first group of questions refers to the actors' attitudes to environmental protection measures: What perceptions do the forest professionals have on the practice of environmental protection in forestry? Could their attitudes explain how far they are prepared to protect the environment? Do the attitudes differ between categories of workers, and if so, how?

Information and education could be presumed to influence people's attitudes. The hypothesis is that the more the individual is informed about environmental protection, and the more he has studied these questions, the more environmental protection measures would he consider in forestry practice. Therefore, questions were put referring to information and education as means for affecting implementing actors' attitudes: What environmental protection education do the forest professionals have? How does it vary between categories of workers? Could information and education explain variation in implementing actors' perceptions of environmental protection?

There are also other ways the forest professionals may be influenced to protect the environment. The discussion in mass-media reports is frequent at times, especially in the newspapers. The forest professionals could be enlightened about environmental issues through membership in environmental associations, for example, nature conservation organisations, ornithological groups, out-door life organisations and hunting associations. In addition, they can be directly contacted by environmentalists who have specific demands on the forest management. Questions that relate to the pressure from environmentalists include: What influence do the forest professionals experience from mass media and environmentalists? Which types of environmental associations are represented among the implementing actors? Do the relations to environmental associations differ between categories of workers, and if so, how?

Since the beginning of the 1950s, starting with the introduction of chain saws, forest operations in Sweden have become almost totally mechanized (Andersson 1982). Today large machinery, such as fellers, harvesters, processors, forwarders, skidders and scarifiers operate on almost all harvestings. This development is generated by economic profit, since the new equipment is much more effective (less labour is needed) than the old. [2] At the same time it has led to a less hazardous work environment for the forest workers.

Commodity values from forests are not generally negatively affected by the introduction of new equipment or methods in forestry (at least not from the point of view of business economics, because then the forestry organisations would supposedly not use them). [3] However, influence on amenity values is not valued in economic terms (except, sometimes, through hunting privileges). Therefore, economic considerations - usually accompanied by a high degree of mechanization - are likely to have a major impact on environmental protection practiced in the field.

The analysis of the field investigation of clearcuttings confirms the hypothesis that a high degree of mechanization negatively affects environmental protection. However, even if the cutting is performed manually, economic considerations remain, for example, as to how many trees could be protected without considerable economic effects to the stumpage value.

In this chapter the following questions are put: How do the forest professionals perceive the effects of mechanization in forestry? How do they explain why highly mechanized harvestings generally achieve less environmental protection? What perceptions do the implementing actors have on the role of economics to the environmental outcome? Could forest economic considerations, according to their view, explain the extent of different environmental features protected? Do the perceptions on harvest technology and forest economic considerations differ between categories of workers, and if so, how?

Perceptions of protection measures

It may be anticipated that the implementing actors' perceptions of the importance of environmental measures correspond to their actual behaviour. As will be described in the following, this is most often the case. There is, however, one outstanding exception, namely protecting nesting trees for birds. Although the saving of nesting trees was valued by many of the respondents, this environmental concern was rarely observed in the field. Evidently, the lack of regard to nesting trees cannot be explained by the implementing actors' attitudes, but probably by other factors such as the chance of detecting them and the design of forest machines.

According to the field investigation, the result of environmental protection depends largely on the choice of harvest technology. Therefore, it is important to analyse whether the differences between manual and machine harvestings are related to differences in attitudes between manual and machine forest workers. For example, the machine workers might have chosen to operate forest machines because they are especially interested in machines, and the manual workers might have chosen to "stay on the ground" because they are more interested in Nature. However, no such differences between the manual and machine forest workers were detected. This indicates that the difference in environmental protection achieved with different harvest operation systems is not, to any large extent, a matter of workers' attitudes but rather the harvesting method as such, as will be further elaborated in this chapter.

The respondents' perceptions on the importance of various protection measures are shown in Table 6.1. Since the question was open-ended a response could focus on either single or multiple protection measures. The aspects of environmental protection described in Table 6.1 are derived from the responses and categorized into seven groups of environmental measures, which may be compared with the results of environmental protection earlier described (See Figure 4.1 and 4.2).

Table 6.1
Importance of protection measures among implementing actors
(Per cent of the category who stated the specific measure as important)

Category of work	Shape of the clear-cut area	Aesthetic values	Do not cut: Non-productive land	Nesting trees	Buffer zones	Wet forest land	Avoid deep tracks from machines	Number of protection measures mentioned per respondent $\bar{x}$	(s)
Forest workers (manual)	5	33	50	61	-	-	11	1.6	(0.8)
Forest workers (machines)	4	23	42	62	4	12	23	1.6	(0.8)
Timber officers	5	45	75	45	30	35	25	2.3	(1.5)
Foresters	9	64	82	55	27	55	-	2.9	(1.3)
Forest owners	11	44	71	44	22	-	22	1.8	(1.6)
Forestry Agency officers	29	63	71	67	46	17	8	3.0	(1.1)
ALL CATEGORIES	12	44	62	57	21	18	16	2.2	(1.3)

Multiple measures were more frequently mentioned by Forestry Agency officers, foresters and timber officers compared to the forest workers and forest owners. The most often remarked "protection" method was that operators should save non-productive forest land [4] (such as bogs and fens and bed-rock areas), which was mentioned by almost

60 per cent of all respondents. Over half of those interviewed valued saving nesting trees for birds. Aesthetic values were also popular concerns. According to field study results, saving non-productive land and aesthetic values are generally among the most observed; however, nesting trees are more seldom retained. In the latter case there is a contradiction between implementing actors' intentions and their actual behaviour, probably due to factors which the actors could not control.

The three categories of environmental protection measures described above - non-productive land, aesthetics and nesting trees - were mentioned to a large extent by all categories of workers. Regarding protection of buffer zones, the answers were more divergent. Almost half of the Forestry Agency personnel stated buffer zones as important, but only one-fourth of the foresters and timber officers - and even fewer from other categories - supported this notion. Wet forest land, on the other hand, was mostly valued by the foresters and some of the timber officers, but comparatively few of the Forestry Agency officers. Both buffer zones and wet forest land were very seldom mentioned by the forest workers and forest owners. The shape of the clearcut area was seldom valued by the respondents; this environmental measure was mostly perceived by the Forestry Agency officers.

Damaged ground was most frequently stated by those categories who operate the machines, the timber officers and the forest owners. Maybe the different categories of workers tend to mention those environmental measures which they perceive as being a problem, which affects themselves, if not regarded. For example, the foresters and timber officers who carry out the detailed planning of the clear-cutting may have observed that cutting small areas of wet forest land is not usually economically motivated. Also, those who operate the machines, or whose forest is affected, perceive deep tracks to be a major problem.

Divergent answers between categories of workers could be explained in at least two ways; first, differences in perceptions of problems, as for the deep tracks, depending on differences in work tasks and, secondly, differences in environmental education. The two explanations are, naturally, not mutually exclusive but rather complementary. The frequency of multiple responses seems to coincide with degree of forest education, which will be further analysed later in this chapter. Protection of buffer zones is an example of a measure which probably is much dependent on education; compared to other environmental features it requires more ecological knowledge, because buffer zones may neither be aesthetic to retain, nor may it be "obvious" that retention of buffer zones increases ecological diversity.

The variation in valuing nesting trees is interesting to

analyse because it did not follow the general pattern. This measure was often mentioned by the forest workers, which might be explained by their work task. The forest workers are able to control the saving of nesting trees more than having influence on all other protection measures, since other environmental measures are usually already determined by pre-harvest planning by superiors. It is important to note the similarity in answers between those who work with chain saws and those who operate machines. They valued almost exactly the same types and number of environmental features.

According to the Forestry Act, the area of each clearcut should be limited. The regulation is vague in this matter, as it only says one should "aim at a limitation of the size". [5] It is interesting to examine how the forest professionals interpret this aim and whether or not they claim to strive at diminishing the sizes of clearcuts. The persons interviewed were asked if they thought there should be a size limit, and if so, what the upper limit should be (See Table 6.2).

Table 6.2
Perceived limits of clearcut area by county
(Number of answers)

Responses	County				Total
	Väster-botten	Väster-norrland	Värmland	Östergötland	
- No limit	5	4	5	3	17
- The sizes should be smaller, but no limit	5	7	2	11	25
- Less than 5 hectares	-	-	-	1	1
- Less than 10-15 hectares	1	-	2	10	13
- Less than 25-30 hectares	1	-	4	4	9
- Less than 50 hectares	15	1	1	1	18
- Less than 100 hectares	6	11	2	1	20
- Less than 200 hectares	1	2	-	-	3
- No answer	-	4	1	4	9
TOTAL	34	29	17	35	115
- Mean size of clear-cutting in the field investigation; hectares	15.5	31.5	6.8	6.6	

There was significant regional variation in answers, corresponding to variation in actual sizes of clearcuts. In the northern counties of Västerbotten and Västernorrland, the clearcuts are generally larger, because the forest land is to a greater degree owned by large-scale forestry (state- and private company-owned) compared to the other counties investigated. In addition, natural conditions and forest history make the northern forests more homogenous - and therefore possible to manage in larger scale. This potential variation in treatment of forests was apparently recognized by the respondents, since the answers in Västerbotten and Västernorrland mostly advocated limits to clearcut sizes around 50-100 hectares, while the respondents from Östergötland (the most southern county investigated) tended to prefer clearcuts smaller than 25-30 hectares.

Almost 40 per cent of those interviewed did not think there should be a limit, although there was a consensus that smaller clearcuttings were preferred. The limit of 40 hectares, which is enforced in Swedish mountainous forests, did not meet an overall approval among the respondents. In Finland, for example, where forest conditions are similar to Sweden, clearcuts are limited to 50 hectares.

When comparing the perceived limits to size of clearcut between different categories of workers, no significant tendencies appear (See Table 6.3). The Forestry Agency officers had as diverse opinions as had the forest operators carrying out the actual harvesting. Apparently, there is no consensus from the Forestry Agency to reduce the sizes of clearcuts. Those who did not want to put up any

Table 6.3
Perceived limits of size of clearcuts among implementing actors
(Number of answers)

Category of work	(n)	No limit/smaller but no limit	<25-30 ha	<200 ha	No answer
Forest workers (manual)	(18)	4	6	8	-
Forest workers (machines)	(27)	10	1	12	4
Timber officers	(23)	9	6	6	2
Foresters	(11)	2	3	6	-
Forest owners	(10)	5	2	2	1
Forestry Agency officers	(24)	11	4	7	2
ALL CATEGORIES	(115)	42	23	41	9

specific limits usually referred to the forest situation (primarily the stand age) which they said "determines the size of the area". Those who wanted to limit clearcut sizes referred both to aesthetic and silvicultural reasons:

The areas should not be too large. Seven to eight hectares is a limit. It is difficult to succeed with replanting if the clearcut area is larger. (Forest worker)

We usually divide the areas so that smaller parts will be cut at different occasions. That depends on the logging operation since we must select solid-ground areas for spring harvesting, because of soggy ground conditions. (Timber officer)

We don't have any problems with the size of clearcuts within this district. But when they become larger than several tens-of-hectares, it does not look nice. (Forestry Agency officer)

The Forestry Act regulations can protect certain flora, fauna and landscape features from being cut. Cutting on non-productive land should be avoided as a rule, but also on "minor productive forest areas of little economic value". [6] There is no definition by the courts as to what is a "minor productive forest area", but the Forestry Agency interprets the limit to 0.5-1.0 hectares (depending on the regional location) in the Forestry Act. [7] The majority of respondents did not want to identify any specific limit (See Table 6.4). A couple of hectares retained without compensation was preferred by a little more than 10 per cent of the respondents, and approximately the same number believed less than one-hectare could be protected without compensation.

Table 6.4
Preferred size of minor forest areas retained for environmental protection among implementing actors (Number of answers)

Category of work	(n)	Nothing could be demanded	A group of trees	Less than 0.5 ha	Less than 1 ha	Some hec-tares	Depends from case to case	Do not know	No answer
Forest workers (manual)	(18)	1	-	1	-	-	5	6	5
Forest workers (machines)	(27)	-	-	-	-	4	6	11	7
Timber officers	(23)	1	-	2	3	4	2	6	5
Foresters	(11)	-	1	1	3	2	1	2	1
Forest owners	(10)	-	1	-	2	3	1	-	2
Forestry Agency officers	(24)	1	2	2	3	1	8	3	4
ALL CATEGORIES	(115)	3	4	6	12	15	23	28	24

There were different opinions between categories of workers;

the foresters and timber officers and forest owners were quite generous to the environmental interests when compared to the Forestry Agency officers. Probably, the opinions of the Forestry Agency officers are influenced by the wording of their own recommendations based on the Forestry Act. Many Forestry Agency officers' answers reflect the case to case judgement:

> It happens very rarely [that a productive area should be saved]. You have to consider the size of the land property.
>
> It depends on the size of the forest. If the Forest Service has a one-hectare stand of young trees on a twenty-hectare cutting, they are usually allowed to cut down the younger trees in the harvest operation. But a small forest-owner is not allowed to do this. Here, there exists no equality before the law.
>
> I don't think we could demand anything - only recommend them to show consideration. Large-scale foresters handle this themselves.
>
> It is difficult because it depends so much on the machine equipment. It is very much a question of negotiations. Most people are reasonable.

One forester perceived industrial forest companies as influential in terms of balancing different interests:

> We have full freedom in our decision. I saved a five-hectare forested peninsula along a lake this spring. (Forester)

Other foresters, however, perceived that decision-making emphasized economic aspects of forest management:

> It depends on the circumstances. But you don't want to diminish the productive area. On less productive land we could however set aside quite large areas. (Forester)
>
> I believe this should not be exaggerated. Totally some single per cent of the clearcut area, especially close to dwellings, should be retained. (Forester)

There were also perceived practicalities for not saving small areas of productive forest:

> It [a small area] is not so easy to save because it often blows down. (Timber officer)
>
> There is no use in saving ten trees at a clearcutting because they will blow down. If the place is protected from wind, then you could save it. (Forest worker)

In summary, the forest professionals' perceptions of the importance of different environmental measures generally corresponded to actual outcome, as shown by the field investigation of clearfellings. Most respondents regarded protection of non-productive land and aesthetic values as important, and these environmental features were also often protected in reality. The only exception was the care for nesting trees, which was frequently valued by the implementing actors although seldom retained in reality. Different categories of workers valued different aspects of environmental protection; the differences coincided with degree of education. Flora and fauna protection measures (except for nesting trees) were seldom mentioned by forest workers; such measures may require more ecological knowledge than, for example, aesthetic environmental protection. There were no significant differences in attitudes between those forest workers who worked manually with chain saws and those who operated forest machines, which indicates that the generally lower degree of environmental protection achieved with mechanized harvesting cannot be explained by differences in forest workers' attitudes.

Role of education

The Forestry Agency, the Forest Service, the private forest companies and the Forest Labour Union have recognized the importance of education and information on environmental protection as ways of changing forestry practice. Consequently, they have initiated an extensive educational programme. During the last five years, numerous courses have been conducted for several thousands of forest workers. In the beginning of the 1980s, when the environmental regulation was fairly new, the majority of courses were held for the Forestry Agency personnel, along with some of the foresters and timber officers from state and private forestry. A series of illustrated books and pamphlets have been produced by researchers on flora and fauna protection in co-operation with the National Board of Forestry. The series also includes a study guide, designed to be an aid during study circles in forest environmental protection.

The analysis shows that the forest workers, who are the ultimate implementors of the protective measures, had studied very little information on environmental protection both during their basic education and after. The forest owners were also little informed compared to the other categories. This indicates a need for more environmental education among forest workers and private forest owners.

During the interviews, the respondents were asked what material they had seen, or read, concerning environmental protection (See Table 6.5). The most widely spread information material was a pamphlet from the Forestry Agency, briefly explaining the content of the environmental protection regulation. The second most read material was the Forestry Act regulation itself. This may be explained in that all the Forestry Agency officers and almost all the foresters and timber officers had read the Forestry Act. The third most spread information was a poster showing fauna protection. The remaining information, including the flora and fauna books, were read by approximately one-fifth of the sample.

Table 6.5
Familiarity with environmental protection information
(Percentage)

Information material	Not seen	Seen, but not read	Read	No answer or not applicable
Forestry Agency pamphlet	38	3	59	-
Fauna protection book	62	16	23	-
Flora protection book	65	14	21	-
Forestry Act regulation	43	8	49	-
Fauna protection poster	55	4	40	1
Fact sheets about endangered species	69	10	22	-
Employer's instruction (if applicable)	4	-	15	82
EPS pamphlet (only in the county of Västerbotten)	17	1	11	70

To compare the different categories of workers, an index was computed both for "information read" and "attitudes" (See Table 6.6). The information index counts "one" for each "seen, but not read" and "two" for each "read" information material listed in Table 6.5. The attitude index counts "one" for each mentioned environmental protection measure in Table 6.1. The Forestry Agency officers rated highest of both information and attitude indexes, which would be expected since most of the information material is distributed by the Forestry Agency. The second most informed were the foresters [8], who also ranked the second on the attitude index, followed by the timber officers. The forest owners were much less informed than the foresters and the timber officers. The forest workers were the least informed; on the average they had **seen** only one information item concerning environmental protection. Out of the total forty-five forest workers, twelve had **read** the Forestry Agency pamphlet, two had **seen** the flora and fauna books, four had **seen** the poster and three had **seen** the Forestry Act regulation. Six of the forest workers had **read** their forest company's own environmental protection instructions. There were no significant differences between the manual and machine forest workers.

Table 6.6
Environmental index score among implementing actors

Category of work	(n)	Information index (0-16) mean	(stand dev)	Attitude index (0-7) mean	(stand dev)
Forest workers (manual)	(18)	0.9	(1.6)	1.6	(0.8)
Forest workers (machine)	(27)	1.4	(1.7)	1.6	(0.8)
Timber officers	(23)	7.3	(3.6)	2.3	(1.5)
Foresters	(11)	6.0	(3.0)	2.9	(1.3)
Forest owners	(10)	3.7	(4.3)	1.8	(1.6)
Forestry Agency officers	(24)	11.2	(1.3)	3.0	(1.1)
ALL CATEGORIES	(115)	5.3	(4.6)	2.2	(1.3)

One factor of relevance to the differences between categories of workers in the amount of information read could be found in comparing their amount of basic education in forestry (See Table 6.7) and their environmental education (See Table 6.8). The forest workers had generally studied very little environmental protection during their basic technical education. No environmental-protection education was offered during their machine and cutting courses. The foresters and timber officers received their environmental education during supervising courses and during their forest technician and forest engineer education, while many of the Forestry Agency officers were educated as forest technicians before any environmental education was included in these studies. More than half of the persons interviewed had not had any basic education in environmental protection.

Table 6.7
Basic forest education among implementing actors
(Number of answers)

Category of work	(n)	No education	Cutting course	Machine course	Supervising course	Upper secondary school (forestry line)	Forest technician	Forest engineer	Master of forestry*
Forest workers (manual)	(18)	2	13	-	-	3	-	-	-
Forest workers (machines)	(27)	6	-	16	-	5	-	-	-
Timber officers	(23)	-	-	1	8	1	10	3	-
Foresters	(11)	-	-	-	1	-	8	2	-
Forest owners	(10)	5	1	-	1	1	2	-	-
Forestry Agency officers	(24)	2	-	-	1	-	21	-	-
ALL CATEGORIES	(115)	15	14	17	11	10	42	5	1

Note: * One person in the category "Others" (Compare Table 2.5) had a master's degree in forestry

Table 6.8
Education in environmental protection among implementing actors
(Number of answers)

Category of work	(n)	Basic education					Further education	
		None	A couple of hours	A few days	More a week	No answer	No	Yes
Forest workers (manual)	(18)	15	1	-	2	-	13	5
Forest workers (machines)	(27)	25	2	-	-	-	22	5
Timber officers	(23)	7	11	3	2	-	10	13
Foresters	(11)	1	7	2	-	1	6	5
Forest owners	(10)	7	-	2	-	1	8	2
Forestry Agency officers	(24)	7	9	5	1	2	1	23
ALL CATEGORIES	(115)	62	31	13	5	4	61	54

Further education had been arranged primarily by the County Forestry Agencies, and almost all of the Forestry Agency officers had attended such courses. The courses ranged from half-a-day to two days, sometimes including a field excursion. Half of the foresters and timber officers had participated, but only a minority of the forest workers and the forest owners. A few of the forest workers said they planned to participate in a forthcoming course arranged by the Workers' Formation Association. [9]

Education and information is the main strategy used by the public and private forest organisations to influence environmental protection behaviour. [10] According to the interviews, there is a positive response to such environmental-protection education. Many of the respondents are eager to learn more (See Table 6.9). The forest workers often pointed out that education should take place during working hours. There were no significant differences between categories of workers in willingness to learn more. Even those who already had participated in environmental protection courses claimed they wanted to deepen their knowledge.

Table 6.9
Desire for more education among implementing actors
(Number of answers)

Category of work	(n)	No	Yes, during working hours	Yes, maybe	Yes, gladly	Conducts courses himself	No answer
Forest workers (manual)	(18)	-	8	1	8	-	1
Forest workers (machines)	(27)	6	8	4	7	-	2
Timer officers	(23)	5	2	5	9	2	-
Foresters	(11)	2	-	1	7	-	1
Forest owners	(10)	2	-	2	6	-	-
Forestry Agency officers	(24)	7	2	4	11	-	-
ALL CATEGORIES	(115)	23	20	17	49	2	4

The relations between education, age, attitudes, information on environmental protection and perceived time deficiencies [11] were analysed using Pearson's product correlation coefficient (See Table 6.10). Because the interviews were made with different categories of workers, respondents were clustered in supervising and forest practice positions with varying educational backgrounds as well as working responsibilities. Therefore, correlations were computed also for three groups of workers: (I) forest workers and forest owners, who generally are little educated; (II) foresters, timber officers and "others", who supervise forest practice; and (III) Forestry Agency officers who carry out the surveillance from the government authority.

With the entire material of interviews, more information read implied that significantly more environmental protection measures were valued according to the attitude index. Also, respondents with higher basic education rated significantly higher on the attitude index. No correlation was found between age and attitudes. However, when the material was grouped, it revealed that the strong correlations between information and attitudes as well as between basic education and attitudes were related to differences between categories of workers. The well-educated groups (II and III) generally rated high on the attitude index and were also well-informed of environmental protection compared to the low-educated group (I) (Compare also Table 6.6).

Generally, the results of Table 6.10 indicate that especially the Forestry Agency officers are a very homogeneous group. Within this group, there were hardly any significant differences in responses. The forest workers and forest owners were the only ones who indicated significant correlations between information/basic education and attitudes. Even when "basic education" was controlled for, those forest workers and forest owners who had been more informed of environmental protection tended to rate higher on the attitude index. [12] Within this group (I) the older respondents tended to value less environmental protection measures according to the attitude index, which possibly could be because they were less educated than their younger colleagues.

Regardless of previous education or age, all categories seemed to want to learn more about environmental protection. Correlations were "to some extent" (at the 10 per cent probability level) significant as to willingness to learn more especially by well-educated forest workers and forest owners (who also rated higher on the attitude index), and well-informed forest practice supervisors (group II).

Table 6.10

Correlations between attitudes, information, education, age, willingness to learn and time for environmental protection (Pearson's r)

	Group	Information index "INFO" (low-high)	Basic education "EDUC" (low-high)	AGE (low-high)	Want to learn more "LEARN" (no-yes,gladly)	Time available for environmental protection "TIME" (little-much)
Attitude index (low-high)		0.44***	0.35***	-0.09	0.11	0.10
	I	0.41**	0.37*	-0.26*	0.33*	-0.11
	II	0.04	-0.05	-0.20	0.01	-0.05
	III	-0.03	0.01	-0.21	-0.11	-0.05
"INFO"			0.75***	0.15(*)	0.14(*)	0.44***
	I		0.36*	-0.05	0.12	0.10
	II		0.42*	-0.19	0.30*	0.28(*)
	III		-0.16	0.49	-0.01	0.28(*)
"EDUC"				0.20*	0.12	0.35***
	I			-0.20(*)	0.25(*)	-0.15
	II			0.01	0.06	0.22(*)
	III			-0.06	-0.20	0.10
AGE					-0.02	0.12
	I				0.06	0.04
	II				-0.20	-0.08
	III				-0.23	0.21
"LEARN"						-0.002
	I					-0.18
	II					0.38*
	III					-0.31(*)

Note: The first figure refers to total n= 115. Missing answers imply that the total n varies, in the table, between 93 and 115

I = Forest workers and forest owners (n = 37 - 52)
II = Foresters, timber officers and "others" (n = 33 - 36)
III = Forestry Agency officers (n = 22 - 24)

(*) Significant at the 10% probability level
* Significant at the 5% probability level
** Significant at the 1% probability level
*** Significant at the 0.1% probability level

Finally, correlations were computed between time available for environmental protection (according to the respondents' experiences) and his age, attitudes and education (See Table 6.10). The strongest influence was found by the factors which measure the respondent's education, such as "information index" and "basic education". The more educated about environmental protection, the more the respondent perceived "time available" as a limiting factor for protecting environmental features. As indicated by the values of correlation coefficients within the groups, this relation was mainly spurious; the persons who are well-educated generally hold higher positions and may experience themselves as having greater responsibility in their work. Consequently, they may perceive themselves torn between multiple tasks, with inadequate time to solve all problems, as was often the case for the Forestry Agency officers and the forest practice supervisors.

Nevertheless, there seemed to exist a direct correlation between basic education/environmental protection information and perceived time deficiences, although it was fairly weak (groups II and III). The hypothesis would be that if the person is well-informed and conscious about the value of environmental protection, he might also find that it takes much time and effort to really consider environmental aspects. He might perceive himself as incapable to detect environmental values in the forest, and feel he should also have more time to influence and follow-up other forest professionals' environmental behaviour. Such direct correlation was, however, contradicted by the lack of influence from the attitude factor (that is the amount of environmental protection measures he had valued important) and the spurious correlation is therefore more likely. The perceptions of time available did not vary according to the respondents' ages.

In summary, the most informed and educated respondents valued more protective measures as being important. The forest workers were very little informed and educated in environmental protection, as were also the forest owners. More than half of the persons interviewed had no environmental-protection education. There were no significant differences between young and old forest professionals in general; thus, the attitudes to environmental protection will not immediately change as an alteration of generations takes place in forestry.

The analysis also indicates that even the persons who were relatively well-educated in environmental protection perceived deficiencies in adequate knowledge and time to pay attention to environmental values. Many of the forest managers and administrators referred to their lack of staff as a major problem if they should practice more environmental protection. But the time factor is closely connected to the information activities and the general attitudes towards environmental protection as expressed by the following citation:

> There is time [enough]. It is more a question of having it [environmental protection] in your mind. I cannot, however, claim this is always done. There also have to be the same thoughts in the minds of the timber officers as of the forest workers. (Timber officer)

Environmentalists

Forestry is a major political issue for many environmental associations in Sweden. Their informatory work and protest campaigns have been intensely followed by mass media (for example, in the 1970s media coverage of demonstrating environmentalists helped stop herbicide treatment of clear-cuttings). Thereby, the environmental movement has probably contributed considerably to the amendment of environmental legislation in forestry and affected the public attitudes towards economic forestry.

Among the many environmental organisations in Sweden, there are several which because of their size and distribution of active members, have been more engaged than others in forest and nature protection issues. The Swedish Society for the Conservation of Nature is the largest [13] and the most established organisation in terms of forest policy programme [14] and contacts with governmental institutions. The Swedish Ornithological Society actively protect bird habitats in the forests. The hunting associations [15] are active in fauna protection and, finally, a number of "outdoor life" and homestead societies [16] often engage in forest management issues when forestry activities affect recreational and social interests in their neighbouring forests. The organisations mentioned above are here generally referred to as "environmentalists".

There are several ways environmentalists attempt to influence the attitudes and behaviour of relevant actors: through mass-media discussions, personal membership in environmental organisations and direct contacts with forest professionals. The analysis of interviews showed that almost two-thirds of the respondents were hunters - the protection of game species was therefore of personal interest to most of the forest professionals. However, membership in nature conservation, outdoor life, and homestead societies was rare. The nature of contacts with such organisations also indicated that the pressure from environmental interests other than hunting was not strong.

The relations with environmentalists were evaluated among the different categories of workers (See Table 6.11). Environmentalists had contacted over half of the Forestry Agency officers, foresters and timber officers. Their degree of contact varied, which is illustrated by the following quotations:

> Somebody sent a message some time ago about an area where there was plenty of aspen trees. I intend to write an "Instruction and Advice" about this to inform the forest owner. (Forestry Agency officer)
>
> I have no steady contacts. Once I was out with a botanist and I ended up writing an "Instruction and Advice" about an orchidé which should be saved. Another time I asked an ornithologist about cutting in an old forest. (Forestry Agency officer)
>
> People from the environmental education at the Folk high school often tell about valuable objects. We have also received a list of capercaille courtship areas from them. (Forestry Agency officer)
>
> When one is cutting close to a village people usually say they want us to leave a buffer. At summer residents we usually ask them how they want us to cut. (Forester)
>
> Some people call us about saving capercaille courtship areas. (Timber officer)
>
> We have had discussions with ornithologists, but only general discussions and no specific objects. (Timber officer)
>
> When we were cutting adjacent to the trotting-course people who lived close to it were worried that the cutting of forest would increase the wind and worsen the climate. (Timber officer)

Some of the respondents revealed through their statements that they personally thought the environmentalists demanded too much from economic forestry:

> The Society for the Conservation of Nature calls me frequently. But they want to save everything - such as entire mountain areas. (Forestry Agency officer)
>
> We had an osprey nest - which was unoccupied - where a guy from the Society for the Conservation of Nature thought we should leave a square kilometre of forest uncut. But we saved one hectare of forest land at another osprey nest where the bird is still breeding. We have an active group of environmentalists here - they often point out special objects. Sometimes we receive letters and this has influenced our acting quite a bit. (Forester)

Table 6.11
Relations with environmentalists among implementing actors
(Number of answers)

Category of work		Contacts						Association membership							
	(n)	None	Hunters	Nature conservation interests	Citizens	A combination of the preceding	No answer	None	Hunting	Nature conservation	Outdoor life	Hunting + Nature conservation	Hunting + Outdoor life	Hunting + Homestead society	No answer
Forest workers (manual)	(18)	18	-	-	-	-	-	9	8	-	-	-	-	-	-
Forest workers (machines)	(27)	22	4	-	-	-	-	14	13	-	-	-	-	-	-
Timber officers	(23)	7	1	5	3	3	5	4	15	-	1	1	-	1	1
Foresters	(11)	3	1	4	1	2	-	-	9	-	-	2	-	-	-
Forest owners	(10)	9	-	-	-	-	1	6	4	-	-	-	-	-	-
Forestry Agency officers	(24)	9	-	11	-	3	1	7	12	2	1	2			
ALL CATEGORIES	(115)	70	6	20	4	8	7	40	63	2	2	5	1	1	1

Among the forest workers, only environmentalists with hunting interests had contacted them. A few said they had discussed capercaille courtship areas. Such areas were paid considerable attention in comparison with other nature conservation issues. The reason is quite natural since so many of the forest professionals expressed an interest in hunting. About two-thirds of the sample interviewed were hunters, and hunting was often referred to as a motive or even sometimes an assurance that environmental protection was considered in forestry.

Few of the respondents were members of environmental organisations other than hunting, although "non-hunting" organisations represent approximately the same national amount of members as the hunting organisations. In terms of memberships, cultural, recreational (other than hunting) and nature conservation interests were thus not adequately represented among the forest professionals.

Some of the persons interviewed had no contact with these organised interests, but still said they were influenced by public opinion and mass-media reports:

> People in general think there are too many clearcuts; but this is more of an overall critique. I could agree on this [too much clearcuts]. (Timber officer)

> Many people dislike [that we are] too much cutting. (Forest worker)

> The more written in the newspapers, the more [trees, wildlife habitat] saved...(Forestry Agency officer)

> We haven't really got any information from the company of what environmental protection is about. But through mass media we have become more conscious. (Forest worker)

Generally, the pressure from environmentalists is not significant according to the interviews. The majority of respondents reported specific contacts with environmentalists in none, or only a few cases. Frequent consultations were very rare. The environmentalists could, however, influence the environmental protection in forestry in other ways; first, through political pressure to legislate environmental protection and during the remiss process, and second, through mass-media attention which may affect the attitudes of forest professionals.

Harvest technology

As previously mentioned, the difference in environmental degradation from mechanized harvest operation systems and manual fellings can be perceived in two ways. One possibility is that the attitudes to or knowledge of environmental protection differs between the forest-machine and chain saw operators. As earlier discussed, the interviews did not indicate such differences. The other possibility is that the design of forest machines, and the organisation of highly mechanized clearcuttings, fail to meet the demands of environmental interests. In the following, it is shown that the interviews confirmed the result of the field investigation.

Manual harvesting was generally considered to be superior to highly mechanized harvest operation systems. A number of reasons of technical and economic nature were suggested by the respondents: difficulties in steering the large equipment, problems in noticing nesting trees from the machines, insufficient pre-harvest marking of environmentally valuable trees and forest areas, demands for a high productivity leading to a fast work pace, and the machines operating during the wet seasons.

The majority of forest professionals said chain saws were better suited than large machines for environmental protection (See Table 6.12). Only a few claimed they had **not** experienced any differences:

> I don't agree... We only use harvesters. I think it is more up to the individual who is operating the machine. (Timber officer)

> Our drivers of harvesters are so interested, and they are older. The manual cuttings are performed by youngsters who are not very interested. They even want to cut down the young stands! (Timber officer)

There were a number of perceived reasons for environmental degradation with mechanized felling (See Table 6.12). No significant differences were shown between categories of workers. The initiative of the driver, and his

Table 6.12
Perceived effects of mechanized harvesting among implementing actors
(Number of answers)

		Harvesting technique best suited for environmental protection				Result of environmental protection at mechanized fellings is largely due to conditions of:					
Category of work	(n)	No difference	Manual sometimes better	Manual always better	No answer	The driver	The machine	Marking of trees	Work pace	Darkness (work at night)	No answer
Forest workers (manual)	(18)	-	4	14	-	3	5	2	1	-	7
Forest workers (machines)	(27)	4	13	4	6	4	9	7	1	1	6
Timber officers	(23)	5	8	6	4	7	7	-	-	-	9
Foresters	(11)	2	6	3	-	5	1	2	1	-	2
Forest owners	(10)	-	1	5	4	-	2	1	3	-	4
Forestry Agency officers	(24)	-	10	8	6	8	3	2	1	-	9
ALL CATEGORIES	(115)	11	42	41	21	27	26	14	8	1	37

concern for environmental protection, was pointed out as a very important factor by over one-third of the respondents. Most of the explanations were of a technical nature. Problems were experienced related to the steering of the felling machine, harvester, or forwarder. However, one of the timber officers claimed there was a difference between the types of machinery. He thought it was easier to turn a new harvester than the felling machines and skidders common in the sixties and seventies. Many also argued it was difficult to discover what trees were appropriate to save during mechanized felling:

> ... The harvester needs a larger space to move. It is easy to push down dead trees and such. (Forest worker)

> Yes, you cut down whatever is in your way. It is so easy to cut with the machine compared to a chain saw. (Forest worker)

> It is difficult because you usually could only see three or four metres up on the stem. But we have probably never really cared about those [nesting] trees. (Forest worker)

> It is easier not to knock dead trees over when you cut manually. The machines are too huge to make precise adjustments. Then you have to leave a broader zone of trees... (Timber officer)

> I don't think there is any difference in discovering wet forest areas, but they have less chance of finding nesting trees from the machines. (Forester)

Many of the answers, especially among the machine operators themselves, put the blame on the detailed planning of the clearcut operation. They thought it would be necessary to mark those trees, or groups of trees, that were supposed to be retained:

> The employer doesn't seem to be interested in saving. The best thing would be if the planning division could mark it out. It is easier for them to see it [the clearcutting] as a whole. Then it would be quite simple to save. (Forest worker)

> It should be marked in advance and the drivers should get more information. That is the greatest problem according to my opinion. (Forest worker)

> Wet forest areas and valuable nesting trees are usually marked when we are cutting our own forests. That is good. Small areas are not retained on our own initiative but only when our supervisors argue about it. (Forest worker)

Because the forest machines are expensive to purchase, they must be highly utilized for profitability. The high demands for utilization may affect the protection of environmental features in at least three ways: the machines work in shifts, which means they operate during the dark period; the operators may perceive a pressure to work as fast as possible; and the machines must work the year round regardless of weather conditions. The work at night was remarked to be a problem by one of the forest workers, who pointed at the difficulty in detecting for example snags and nesting trees in the dark, but another of the forest workers did not agree:

> We have plenty of light [on our machines] and during our break, which lasts for three hours between the working shifts, we go out into the forest and look. If you are interested in moving about [in the forest], then you have enough time. Here, all the operators of forest machines are former manual forest workers and interested in forestry.

The work pace was experienced as the main problem by a total of eight respondents:

> It is more bustling to work with the machines. (Forest worker)

> ...It is altogether a matter of productivity. We need to diminish the demands for productivity of the machines. (Forester)

> Usually we are in less contact with the machine drivers. And they are working at piece-rates. The manual workers have a fixed salary and are not as forced [by time]. And there are also the entrepreneurs who drive the machines and cut down everything - ...(Forester)

Deep tracks emerging from forest machines were perceived as being an environmental problem by a majority of respondents among all categories of workers, including the machine operators (See Table 6.13).

Table 6.13
Perceived impact of machine tracks among implementing actors
(Number of answers)

Category of work	(n)	No problem	Quite a problem	A big problem	No answer
Forest workers (manual)	(18)	-	7	-	11
Forest workers (machine)	(27)	1	15	5	6
Timber officers	(23)	-	9	4	10
Foresters	(11)	-	5	3	3
Forest owners	(10)	-	3	2	5
Forestry Agency officers	(24)	2	6	7	9
ALL CATEGORIES	(115)	3	47	21	44

Damage to the ground through deep tracks is not only visually obtrusive, but could negatively affect both remaining vegetation and soil conditions, for example through erosion. Therefore, the Forestry Act requires restoration of damaged ground in forest areas that are intensively used for recreation, as well as damaged forest paths and roads. Nevertheless, damage caused by too heavy machinery is common:

> It is terrible the way it looks at times. I think the damage from deep tracks is one of the greatest problems in forestry of today. (Forest owner)

> It is almost impossible to manage. We have to fill in too deep tracks with excavators. Everyone violate the Forestry Act when it comes to deep tracks. (Timber officer)

> It is a big problem, and will be very difficult to avoid. You cannot cancel the loggings when the tracks appear - who should pay for refilling the tracks? (Timber officer)

Prevention of damage is often costly because it requires short-time changes in harvest planning. Operations with large machinery on soft-grounded forest land must be carried out when the ground is solid with frost, and during rainy periods transports on soft ground must be avoided. Most of the deep tracks are caused by the forwarder, which transports the wood. The harvester often works on a bed of logging residues, where the pressure on the ground is more evenly spread. The forwarder, however, uses the same path over and over again. The larger the clearcut area, the more wood must be transported along a single main haul road, and the more damage is made.

> It has become more and more difficult to find hard-grounded areas to harvest during the breaking of the frost...(Forester)

> ...We don't have any real winters here - it is very rarely ground frost. The landscape is hilly, with much wet ground in between still you have to pass there [with the machines]. (Forester)

> It is difficult to plan. Sometimes the winter is bad [warm], and the machines are too heavy. Here, we only have big, heavy machines. You should have less heavy machines - maybe they will arrive. (Forest worker)

If the forwarders were made smaller, the problem would - according to another forest worker - still exist:

> Smaller machines are coming, but then the problem is they carry too small a load...(Forest worker)

Maybe, the best solution until more flexible forest machines are utilized was described as follows:

> During the breaking of the frost we usually put the machines aside for a couple of weeks and use the time for education and repair of the equipment - to avoid damage. (Forest worker)

The Forestry Agency officers did not think it was possible to actually prohibit clearcutting because of the risk of damage to the ground. However, quite a few of them had required restoration of deep tracks after the cutting was finished.

> Of course this is a problem. But what could we do - we cannot stop any cuttings because of this. The moving of

> machines [from one clearcut area to another] costs a whole lot of money. (Forestry Agency officer)

> It is a terrible problem, especially at the hillsides where the water flow could change and landslides might occur. The last couple of years some repairs of the worst damages have started. Forest roads are also damaged, impeding small tractors to drive in the deep tracks. (Forestry Agency officer)

> This [deep tracks] is a bigger problem, as I see it. Some entrepreneurs have operated at the wrong season. We have then required repairing the damage. (Forestry Agency officer)

Forest economic considerations

There is a fundamental conflict between removal of as much timber as possible from a forest area and protecting the environment, since the latter usually implies retaining a certain amount of timber. As earlier described (See Table 6.4), most forest professionals did not advocate protection of large, productive forest areas for environmental purposes without economic compensation. In addition, the majority did not think there should be a specific limit to the size of clearcuts (See Table 6.3), although most of them perceived very large clearcuts to be unnecessary. Corresponding attitudes were found with the attitudes toward deep tracks. Although deep tracks were generally regarded to be obtrusive, most respondents thought it was impossible to avoid ground damage. The piece-rate system for paying forest workers and the lack of economic financial support to compensate forest owners were regarded to be important economic constraints to environmental protection.

The conflict between economic and environmental considerations was perceived differently (See Table 6.14).

Table 6.14
Perceived role of economics and conflicts among implementing actors
(Number of answers)

Category of work	(n)	Role of economics: No importance	Some importance	Great importance	No answer	Conflicts: Do not exist	Exist; easy to solve	Exist; sometimes difficult to solve	Very difficult to solve	Conflicts within the legislation	No answer
Forest workers (manual)	(18)	-	2	11	5	-	12	6	-	-	-
Forest workers (machines)	(27)	2	6	-	19	14	2	5	-	-	6
Timber officers	(23)	5	7	4	7	2	9	6	3	-	3
Foresters	(11)	3	1	2	5	4	3	3	-	1	-
Forest owners	(10)	2	3	2	3	-	2	4	-	3	1
Forestry Agency officers	(24)	1	7	10	6	2	11	11	-	-	-
ALL CATEGORIES	(115)	13	26	31	45	22	39	37	3	4	10

Many respondents, especially the forest workers, did not express any clear opinion about the role of economic considerations. Few respondents, however, claimed economics did **not** influence the degree of environmental protection. There was a difference between those forest workers who operated machines and those who used chain saws. The manual workers seemed to believe economics was a more important factor, and perceived more frequent conflicts between economic forestry and nature preservation than did the machine workers. It is difficult to evaluate the significance of the different views because many of the machine workers did not answer. It may be that conflicts are so common, and the role of economics so dominant with mechanized harvesting, that the forest workers operating the machines have become too used to the conflict to be able to recognize it. Two of the forest workers who claimed economics play a major role expressed their opinions in the following way:

> When it comes to the sizes of clearcuts they [our supervisors] blame economic reasons for not being able to reduce the sizes. Money rules everything. The forest machines work at piece-rate and it is difficult then to have the time to consider environmental interests. (Forest worker)

> It must be possible to combine [economic forestry and environmental protection]. But nowadays everything is economics. If you could decide things yourself you might have saved more (trees), but you cannot say money rules <u>entirely</u> - that is an exaggeration. They always kept on at me about cutting everything down, so now you don't see what could be protected. You just look at the forest in terms of cutting down. (Forest worker)

Many of the respondents thought the piece-rate system in forestry was an important constraint to have time for environmental protection. When the forest workers are paid piece-rate, there is an economic incentive for them to cut everything. The design of wage contracts for forest workers varies between different parts of Sweden and between different employers. The interview study did not specifically investigate differences in wage construction between manual and machine workers. However, many of the machine operators said they felt obligated to keep the machines going - they had a clear understanding of the financial losses if the machines were not constantly used. Therefore, most machine operators never walked through the forest before harvesting to detect any environmental features. Protection measures must therefore be marked in advance, for example, by the timber officer. Some of the machine operators who own their equipment perceived even more economic pressure to work as fast as possible than those who do not own their equipment. However, they also felt a pressure to satisfy the demands of the forest owner in order to maintain their work and reputation.

Economic aspects were discussed in the small scale of leaving a couple of trees, or a buffer along a stream, and in the large scale such as retaining several hectares of forest land.

> Of course there is some economic sacrifice in leaving behind ten big pine trees. But after taxes are paid, there is not that much [money] left...(Timber officer)

> There has been a discussion sometimes even with the forest workers. But if you are planning well, there is no great cost of leaving vegetation for example along a stream. It is a matter of planning, since you cannot guard every tree. (Timber officer)

> Yes, it [conflict] happens. For example, it is often too expensive to come back and harvest seed trees at a later occasion. [17] It ends up being the wallet of the forest owner steering things. (Timber officer)

> More and more large forest areas will be proposed as reserves for rare plants. Forestry will suffer from this. So far, this is only a discussion - we haven't had a concrete case. (Timber officer)

The officers from the Forestry Agency experienced more conflicts than other categories of workers. Often these cases of conflict reflect the limited state fundings for economic compensation to land owners. When the environmental protection authorities cannot compensate for the loss of land, the Environmental Protection Section at the County Administration has no more authority than the Forestry Agency to enforce environmental protection.

> There was a virgin forest of five hectares which we intended to fully retain. But the owner was a private individual, and the National Environmental Protection Board had no money to pay for it. I thought this was terrible because we won't find these kinds of forest any more. Now there are only seven or eight pine trees left - everything else was cut down. I think this was a strange turn. The NEPB said they were informed about it too late - but it has been known to environmentalists since the sixties. (Forestry Agency officer)

> There have been areas of valuable forests with rare plants, but we have no money... When money is lacking the Environmental Protection Section has let go and not claimed to protect the area... There is another area which the forest owner agreed to save, but there the terrain is rough and on the verge of productive land. (Forestry Agency officer)

> Around here, there are plenty of orchids, for example lady's slipper. But you cannot prevent a forest owner from cutting. And when they have clearcut you cannot prevent them from draining the area to be able to regenerate. Only draining of forested areas is possible to prevent by legislation. [Hence] You can get around legislation. (Forestry Agency officer)

Draining of forest land must be approved by the EPS authorities, which on the basis of environmental inventories can prohibit draining if it will destroy unusual flora and fauna habitats. However, draining clearcut areas is allowed without any permission. Draining a clearcut is called "protectional draining" in Swedish because the aim is to protect the growth of seedlings, which otherwise would be threatened by the surplus of water. Supposedly, the effects of the "protectional draining" should not last for more than a couple of decades and not imply a lasting change of site conditions. Nevertheless, it may affect flora and fauna which depend on wet forest land. Since draining of both forested lands and clearcuts have largely increased during the last decade in Sweden, the environmental effects of such draining could be quite extensive.

*

In summary, respondents frequently cited harvest technology and forest economic considerations as important factors affecting the result of environmental protection. The forest machines that were used in mechanized harvestings were considered inadequately suited for protecting environmental features in several ways. They may be difficult to steer when small-scale considerations are demanded (such as leaving single trees); nesting trees are not easily detected from the harvesters which makes pre-harvest marking of such trees necessary, and this pre-harvest marking is not often carried out; in addition, the machines frequently cause deep tracks, especially the forwarder when too much wood has to be transported along the same main haul road.

Demands for high economic profit put pressure on the machine operators to work at a fast pace, which makes environmental considerations difficult to manage: the forest workers are often payed piece-rate, the machines operate all year round (often at dark) and regardless of weather. The high productivity demands on forest machines, caused by the high investment costs, thus work against environmental protection. Enough time may not be allocated to mark out environmentally valuable trees and forest areas in advance and the operation may not be cancelled if environmental damage is expected, for example because of weather conditions.

Economic conflicts are perceived both in the small-scale of retaining single trees and in larger-scale, such as protecting several hectares of forest land. Most of the small environmental measures could, however, be carried out without economic conflicts if they are adequately planned for (except, maybe, for some small-forest owners who perceive great value in single trees).

The lack of resources for compensating forest owners for their economic loss when several hectares should be protected is a big problem, especially for private, individual forest owners who do not generally own large holdings and therefore would be more economically affected by a prohibition to harvest a certain forest area. Many of the environmentally most valuable forest areas cannot be protected since there are no state fundings for preserving such small areas. As it is now, the environmental authorities have to give priority to large areas with outstanding environmental values, considering the amount of resources for nature preservation. If economic incentives could be provided from the State to forest owners for protecting forest areas, or manage them with special caution, more environmental protection would most likely be achieved.

One of the Forestry Agency officers explained the effects

of harvest technology and forest economic considerations in a way which summarizes what many others also expressed:

> Forest owners should be able to get reduced taxes if they are to set aside large forest areas for nature conservation or cultural purposes. Usually, however, we can work out sensible solutions. The economic demands have increased with the mechanized harvestings. (Forestry Agency officer)

NOTES

[1] The exact questions that were discussed during the interviews are presented in Appendix 2.

[2] The difference in labour needed for different harvest technology is illustrated by Andersson (1980, p.69): Clearcutting with chain saws in a Swedish conventional shortwood system requires 11 day's work per 100 cubic metres harvested, while clearcutting with a harvester requires approximately 3 day's work for the same volume of timber harvested.

[3] For example, harvest costs can be reduced by 21-35 per cent if harvest technology is changed from clearcutting with chain saws in a shortwood system to clearcutting with chain saws and processors (Forestry in the Long-Term Planning Commission, Ds E 1980:4, p.135). Generally, the choice of harvest technology depends on a number of factors, which vary according to local conditions, such as conditions of the terrain, available labour and machinery etc.

[4] As earlier discussed, such areas are retained primarily of economic and technical reasons.

[5] Forestry Act § 21 prescriptions by the National Board of Forestry.

[6] Forestry Act § 21 recommendations by the National Board of Forestry.

[7] Ibid.

[8] In Table 6.6 the category "others" were actually the second best informed, but the figure is disregarded in the analysis because the category only consisted of two respondents.

[9] In Swedish: Arbetarnas Bildningsförbund (ABF), an association which arranges study circles in cooperation with the Labour Union.

[10] Both the Forestry Agency (Compare Chapter 7), large-scale forest owners, and the Forest Workers' Labour

Union emphasize education and information as means for influencing environmental considerations in forest management. For example, a large number of pamphlets on environmental protection have been produced, and excursions, seminars and courses arranged (Compare Chapter 1).

[11] In Chapter 7 perceived time deficiencies for protecting the environment are further analysed.

[12] With multiple regression the influence of "basic education" and "information" on "attitudes" for group I were analysed. The beta-coefficients were 0.17 (T=1.28, thus insignificant) for "basic education", and 0.55 (T=4.13, significant) for "information" (total r=0.63).

[13] In 1987 their membership was around 130,000 members in 241 local groups (oral communication with their headquarters in Stockholm). In addition their youth organisation is very active. The number of members might today be exceeded by Greenpeace, which is a new organisation but strongly growing in Sweden. However, Greenpeace has not been very active in forest issues.

[14] See, for example, Olsson (1985).

[15] The two hunting associations (Svenska Jägarförbundet and Jägarnas Riksförbund/Landsbygdens Jägare) count around 170,000 members. (Oral communication with their headquarters.)

[16] Homestead societies engage in preservation of cultural resources and are locally active in protecting old settlements in the forest.

[17] What he meant was that regeneration by seed trees is often preferred for environmental reasons (the seed trees mitigate the clearcut impression), but that the seed trees must be cut later on, when the new generation of seedlings is established.

7 Steering environmental protection through legislation?

In the top-down perspective, there are five factors which may be exerted from implementation literature as being important to analyse in explaining implementation outcome. These factors are previously described in Chapter 1 and include: (1) the form and content of the policy itself; (2) the capacity of the formal organisation(s) responsible for making the programme work; (3) the qualifications of the people in charge of the operations - "the will, skill and attitudes of implementors"; (4) the environment, to which socio-economic conditions, technology, media attention, public support, attitudes and resources of constituency groups (e.g. environmentalists) are alluded; and (5) the influence of time.

Factor (1) includes not only the clarity, specificity and/or flexibility of goals and procedures, but also the strategy with which the implementing agencies are to work with the programme. Resources - financial and others- together with implementing-agency organisation and culture constitute factor (2), while attitudes and perceptions of implementing personnel, multi-actor structure and number of actors refer to factor (3). General societal conditions, which may affect the implementation of the programme include the technical conditions for preserving environmental features as well as pressure for policy changes from environmental groups. Timing is important, since attitude changes are likely to occur gradually over a long period. Previous implementation studies have often neglected the influence of time, and therefore come to more negative conclusions about the effect of a certain policy than if it had been evaluated over a long period (Mazmanian and Sabatier 1983; Sabatier 1986).

In this chapter, the effects of legislation on environmental protection in forestry is examined. Steering is assumed to be exercised by the government and its agencies. It is time now to discuss the Swedish forest-environmental regulation as a method for steering. The analysis focuses on methods of how the behaviour of implementing actors is influenced by the legislation, and on how the forest professionals' actions for protecting environmental features may be controlled. More precisely, how does the Forestry Agency make use of the regulation? What actors control the implementation process: the government, the bureaucrats and the experts, in the hierarchical, representative democratic tradition or the "ordinary" citizens in a grass-root democracy? Possibly, the frame-law design of modern environmental laws may even require some form of public participation in the implementation process. As Ericsson (1985) points out, the system of frame-laws makes it necessary for the rule-making agency to communicate with public interests, because the administrating agency must take a political stand.

The analysis in this chapter is based on comparison with other similar studies regarding the effects of legislation and regulation on policy implementation. It includes an investigation of how the Forestry Agency uses prescriptions and advice in order to influence forest owners to take environmental measures. Moreover, a comparison is made between forest-environmental legislation in the United States and Sweden, with the aim of introducing and comparing alternative strategies to cope with the environmental problem. A study of the impacts of the Forest Practices Act in the state of California is analysed, because its requirements for environmental protection are strong, and contrasted to the Swedish study. The comparison suggests ways that the forest professionals' awareness of environmental values may be increased, and concludes that significant environmental protection cannot be achieved without economic input, such as expenses for pre-harvest planning and review.

Swedish forest legislation as a steering instrument

Generally, influencing behaviour may be classified along two dimensions: (1) as normative, economic or coercive and (2) as rewards or punishments. Normative inducements are those involving "esteem, prestige and ritualistic symbols" and "acceptance and positive response". Economic inducements involve material resources. Coercive inducements involve "infliction of pain, deformity, or death" and "restriction of movement" (Etzioni 1975).

Legislation is often thought of as binding decisions accompanied by legal sanctions to be used if the requirements are not fulfilled. Recent legislation, however, tends to be of frame-work character (Hydén 1984). Environmental legislation usually includes many statements of regulatory character. Although coercive elements may be (a minor) part of the regulation, their nature is nevertheless mostly non-coercive. If such statements included in the legislation are to be influential on people's behaviour, it is crucial that their content is communicated to the people affected, and that it is understood by them. In addition, the target group who is to comply with the regulation must be motivated to do so. A general respect for legislation is not always enough, but the compliance must generally appear favourable. Often the regulation confirms an already existing moral code. The regulation may also assist in changing attitudes (Eckhoff 1983).

Although the environmental regulations of the Forestry Act are accompanied by a possibility of legal punishment, such coercive inducements are very rarely utilized by the Forestry Agency. Their aspiration is that the forest owners follow the environmental recommendations naturally without any prodding from the authorities (Falk and Mörtnäs 1984). The influence on the behaviour of forest owners and those who work in the woods is almost exclusively exerted by normative inducements - information, advice and persuasion. The environmental regulation is thus assumed to be congruent with already existing social customs and ways of thinking.

A hypothesis is that the law only confirms norms that the supervising Forestry Agency, environmental authorities, and the target groups already have. Stjernquist (1973) examines the impact of the Swedish Forestry Act on small-forest owners' behaviour during 1905-1960. He also analyses the strategies with which the Forestry Agency officers use the regulation. During the whole period investigated there was a clear parallel between the development of legislation and the development of improved forest practices in private forests.

However, this does not prove that the laws were the decisive cause of the changes in behaviour. Other factors were analysed by Stjernquist, which may have worked simultaneously with a greater effect than the law itself: (1) political and social pressure to change the behaviour of forest owners; (2) social and technical changes or changes in the structure of forestry holdings; and (3) economic improvements in the forest owners' situation. Of these factors, only the last one was found to have had a significant impact, since the forests' economic value increased from the end of the 1930s. Political and social pressure to change the attitudes was not applied until the 1950s, and major social and technical changes were not found during the period investigated. Stjernquist (1973) concludes that the activities of the legal authorities were the principal reason for changes in behaviour.

Nevertheless, he does not attribute the change in behaviour to the regulation alone. Instead, the major impact is ascribed to the strategy of give-and-take methods and mutual understanding between the field officers of the Forestry Agency and the forest owners. The law may also have had importance as a kind of authorization for the field personnel. Three circumstances made the forestry programme effective in the field:

> (1) No organized resistance. This does not mean, however, that the programme as such was accepted by the forest owners. There were several obstacles in the way: firstly, that the contents of the programme were vague and incomprehensible and secondly, that the programme was dressed in legal terms and therefore became compulsory. The legal form did not give the programme legitimacy but on the contrary impeded acceptance.
> (2) Good contact between the authorities and the forest owners. The programme was not accepted in its legal form, admittedly, but in the more concrete form it was given in the appeals and arguments by the field personnel. One condition necessary for this seems to have been that the law and legal sanctions were not invoked. The field personnel had to work with social pressures, particularly different kinds of reciprocity.
> (3) Financial opportunities for the forest owners to follow the programme. The measures requested did not involve any great cash outlay, mostly the owner's own work. In addition, the county forestry boards made contributions to the costs to a great extent (Stjernquist 1973, p.205, emphasis mine).

A comparison between Stjernquist's study and the study of the implementation of environmental protection regulation reveals both similarities and differences. The environmental protection regulation is vague. It leaves room for different interpretations and bargaining at the field level between the authorities and the target groups. Also, the Forestry Agency use of the regulation is characterized by soft steering, with friendly advice rather than rigid rule enforcement.

In Stjernquist's study the attitudes of Forestry Agency officers corresponded to the Forestry Act requirements. Hence, the Forestry Agency officers often do not even refer to the law when convincing the forest professionals how to behave, but rather use personal motivations (Ibid., pp.150-3).

In the case of environmental protection, however, the interviews indicated that the Forestry Agency officers' attitudes seldom give priority to environmental protection compared to other, competing, requirements according to the Forestry Act. They generally perceived their resources inadequate to be able to detect environmental features before forest operations are allowed. Hence, specific advice to forest owners concerning a particular clearcutting were rare; instead, the Forestry Agency as a whole works primarily with environmental-protection information for forest owners and operators. The handbooks, pamphlets, and the study programmes that the NBF have initiated stand out as exemplary means for influencing the forest professionals' attitudes toward environmental protection in general.

The financial inducements differ in the two studies. In Stjernquist's study economic inducements worked hand in hand with the Forestry Act requirements. Since protection of environmental features often implies a certain economic loss of wood, forest economic considerations (to varying extent in different cases) counteract environmental protection. In general, economic values of non-timber resources are seldom taken into account by the forest professionals, and the Forestry Agency does not give subsidies to forest owners who retain significant forest areas for environmental reasons. On the contrary, the tax system may even restrict how much forest they can protect. This is because taxes are based on the commercial value of their forest holding, whether or not the entire forest can be used for timber production alone. Forest areas that should be protected for environmental reasons are therefore still taxed.

Forestry Agency officers' conditions of work

The District Forest officers of the County Forestry Agencies are the ones responsible for monitoring the Forestry Act regulations and giving advice to forest owners. They are to make the case-to-case judgements of what environmental protection is "appropriate" according to the legislation. Their influence on forest professionals' behaviour can be exercised through prescriptions and advice. With Lipsky's (1980) terminology, these officers may be classified as "street-level bureaucrats". The street-level bureaucrats work at jobs characterized by relatively high degree of discretion and regular interaction with citizens. Their work is further characterized by: (1) chronically inadequate resources relative to the tasks workers are asked to perform; (2) a demand for services that tends to increase; (3) goal expectations that tend to be ambiguous, vague or conflicting; (4) performance oriented toward goal achievement that tends to be difficult to measure; and (5) clients who are typically non-voluntary. As a result, clients for the most part do not serve as primary reference groups (Lipsky 1980, pp.27-8).

With the exception of the statement about non-voluntary clients, Lipsky's characteristics apply well to the District Forest officers. As the majority of DFOs pointed out during the interviews, resources were inadequate if environmental protection features were to be detected and the forest owners sufficiently informed:

> The forest owners often lack knowledge. If we had time to check-up on more clearcuttings the environmental protection would increase substantially. (Forestry Agency officer)

> This is a dilemma. There is little time to devote to environmental protection. (Forestry Agency officer)

Demands for environmental protection advice are not significantly increasing. However, other requirements of the Forestry Act are growing: stricter control on insect damages, compulsory pre-commercial thinning of young stands, re-establishment of forest stands which do not adequately use the soil fertility, and production of Forest Management Plans. Consequently, the Forestry Agency officers are faced with an abundance of different tasks. The total workload of the DFOs is significant, and many of the DFOs feel they have too little time to devote to environmental protection issues. Still, the lack of environmental protection control cannot be simply explained by differences in workload. Rather, it is the attitudes of the individual DFO that explains how he gives priority to different tasks:

> Actually this [environmental protection advice] does not take much time. You walk around there anyway. It is the knowledge of environmental protection that is not good enough. (Forestry Agency officer)

Similar findings are made in a Danish study of a social welfare reform; even though the actual and perceived workload per social worker varies considerably among the local welfare agencies, there is no correlation at all between the two variables; more important, the actual workload does not explain variation in the social workers' actual capacity to reach their clients (Winter 1984). Instead, coping behaviour is closely related to perceptions and attitudes, which emphasizes the importance of organisational culture.

As earlier described (See Chapter 3), Forestry Agency officers' goal expectations are vague and often conflicting, due to remaining conflicts between timber production and environmental protection in the Forestry Act. In addition, the Forestry Agency has a split role in relation to the forest owners, since their commercial activities may restrain their surveillance activities. As a consequence of the vague guidelines, goal achievements of environmental protection are difficult to measure. The judgement of the extent of environmental protection required is made from case to case, and there are no standardized measures by which environmental achievements may be evaluated.

In Lundberg's (1982) study of the implementation of the

Swedish Work Environment Act, he finds the street-level bureaucrats - the inspectors - torn between the role of policeman and the role of advisor. These two roles cannot both be exercised in a satisfactory way at the same time, because they are contractictory. Since their organisational culture favours the role of the friendly counsellor, the inspectors choose to avoid legalistic means for influencing employers to improve the work environment. The inspectors adapt themselves to economic interests (Lundberg 1982, pp.150-2). As showed in the following, the Forestry Agency officers' use of formal prescriptions and advice is also characterized by the soft steering. Punishment has so far never been used as a mean for controlling forest owners' environmental protection behaviour.

Forestry Agency use of prescriptions and advice

In addition to general information activities, the Forestry Agency officers may influence the forest owners specifically to protect environmental features. These instructions to the forest owners are based on the Forestry Act regulation. Therefore, their content reflects the way the regulation is interpreted and implemented by the Forestry Agency. Prescriptions (which if neglected may lead to fine) are very rarely issued. Instructions and advice are more common, but cannot be legally enforced.

In order to evaluate the use of formal prescriptions and advice as means for steering behaviour, these documents were examined. All 12 prescriptions and 842 written advice documents (which represent 35 per cent of the total registered written advice) issued by the Forestry Agencies in all Swedish counties from 1980 through 1986 were investigated. First, the content of the documents was analysed in comparison with the environmental regulation: What aspects of environmental protection were mentioned in the documents? To what extent were environmental measures required? Second, differences between counties and districts were investigated: Did regional differences exist in the use of prescriptions and advice, and if so, how? Did different age groups of Forestry Agency officers use the formal advice differently, and if so, how? Third, the extent of contacts with government and private organisations was examined: How often were the environmental authorities, municipalities or environmentalists consulted? Were there any differences in the use of prescriptions and advice depending on which forest owner category is addressed? Fourth, the point of time when the document was issued was examined: Had the use and content of formal instructions and advice changed during these six years, and if so, how? Finally, the season when the advise was given was examined in order to look for problems in detecting environmental features during winter time.

Content

The instructions and advice contained elements of environmental protection from both aesthetic/recreational and biological/ecological aspects. Protection of non-productive forest land [1] (9 per cent) and aesthetic trees (9 per cent) were specified. Examples of recreational measures included restoration of deep tracks and clearing of logging debris (13 per cent). The majority of biological/ecological environmental measures required in these documents comprised buffer zones (22 per cent) and snags and nesting trees (15 per cent).

Of the prescriptions, five were issued because of recreational interests [2]; the remaining seven prescriptions were motivated by biological reasons and demanded protection of a variety of different environmental features. [3]

Still, the high degree of environmental protection achieved of aesthetic and economic/technical reasons according to the field investigation (See Chapter 4) does not seem to be a result of the content of Forestry Agency advice; if so, more emphasis on these values from the Forestry Agency would have been expected.

Very few of the protection measures specified were historic sites (3 per cent of the instructions and advice and none of the prescriptions). Furthermore, specific floristic values were rarely mentioned. Only three of the examined instructions and advice were specifically issued with the intention of protecting valuable flora habitats, as was also one of the prescriptions. Thus, it is believed the Forestry Agency personnel are insufficiently educated in protection of cultural resources and rare and threatened flora species. A handbook on cultural protection in forestry is under production by the National Board of Forestry in collaboration with the Central Board of National Antiquities (in the same series as the fauna and flora protection handbooks), which indicates that the government authorities are aware of this problem.

The extent of environmental protection required by the instructions and advice ranged from single trees to, at the most, approximately ten hectares. The large areas were primarily non-productive land, forest areas adjacent to dwellings and breeding/courtship areas. On these areas, the advice given from the Forestry Agency was to retain (for example) aesthetic trees and do selective cutting/ thinning instead of total clearcutting.

With a few exceptions, the content of the prescriptions and written advice generally reflected the variety of environmental measures required by the Forestry Act. On the whole, the demands on environmental protection from the forest authority were very moderate. The majority of documents required much less than half-a-hectare per clearcut retained for environmental reasons.

Differences between regions and officers

All County Forestry Agencies used written instructions and advice to some extent, but only seven of them had issued prescriptions. The amount of written instructions and advice examined in the investigation varied from one in the county of Västmanland to 234 in the county of Västerbotten. Västerbotten accounted for almost 30 per cent of all investigated instructions and advice. The county of Kopparberg had issued the most prescriptions.

Divergent use of written advice does not necessarily reflect actual differences in advice given to forest owners because the following two circumstances may vary between the counties: first, all documents issued may not have been saved and sent in for the analysis and second, advice may be oral.

Generally, the content of the instructions and advice issued in the different counties varied according to divergent natural conditions. Recreational interests were, for example, most frequently mentioned in the southern parts of the country. Two environmental measures, however, were totally lacking in some counties; in fourteen counties environmentally rich areas, such as marshes and ravines, were never mentioned and in eight counties breeding and courtship areas were totally neglected. The absence of these environmental measures in some counties might simply be a result of the small amount of environmental advice issued.

Some Forestry Agency officers were more active than others - for example one District Forest officer alone had issued 147 written instructions and advice, which corresponds to 18 per cent of the total material investigated. His documents contained mostly general environmental advice, which indicates that he had not specifically investigated what environmental features were present at the proposed clearcuts.

During the interviews with Forestry Agency personnel, some of the officers expressed a feeling that younger Forestry Agency officers were generally more active than older officers in putting pressure on forest owners. Furthermore, some of the Forestry Agency officers believed that the younger officers were more educated in environmental protection and brought up with a "new" image of the Forestry Agency, where surveillance of the compliance to the Forestry Act was more emphasized than before. This hypothesis was neither supported by the interviews on environmental education (Compare Table 6.10), nor by the analysis of written instructions and advice. On the contrary, the older District Forest officers showed to be more active than the younger. Those born in the 1920s had issued 4.8 documents on an average, while those born in the 1950s had written 2.9 each. The rest of the DFOs were in between. The content of the instructions and advice issued did not vary significantly between the DFOs.

Contacts with public and private organisations

According to the investigated written instructions and advice, the Forestry Agency consulted other government authorities and/or spokespersons for environmentalists and local interests in approximately 20 per cent of the cases. Most of the consultations were made with the EPS at the County Administration (14 per cent); the municipality was more seldom conferred with (5 per cent) and environmentalists and local individuals rarely asked (1 per cent). Of course, the consultations documented on the written advice on environmental protection are only a part of all discussions carried out between the Forestry Agency and other organisations. Still, the result indicates that the municipalities and the public, impacted by forest operations, are seldom consulted.

Half of the prescriptions and 80 per cent of the written instructions and advice were directed towards private, individual forest owners. Private forest companies were addressed in few of the formal documents (16 per cent of the prescriptions and 7 per cent of the advice) compared to their share of forest land. One third of the prescriptions and 13 per cent of the advice were directed towards state- and commonly-owned forests. This indicates that the Forestry Agency officers tend to pay more attention to the forest activities of private individuals than private companies, despite the private companies' lower degree of environmental protection (Compare Chapter 4, See Figure 4.4). The extent of environmental protection required in the documents, however, was the largest for the private companies compared to the other forest owner categories. Thus it appears that the small amount of advice given to private companies is, to some extent, compensated by greater demand to protect comparatively large forest areas for environmental reasons.

Timing

The frequency of written prescriptions and advice has gradually increased since the amendment of the environmental regulation in 1980 (See Table 7.1). No significant differences were found with regard to the content of environmental measures required over these six years. Hence, nothing indicated a change of attitudes among Forestry Agency personnel in terms of what aspects of environmental protection were emphasized.

Table 7.1
Number of "instructions and advice" and "prescriptions" during 1980-1986

Year	Instructions and advice	Prescriptions
1980/81	83	-
1981/82	169	1
1982 (half-year)	73	-
1983	371	3
1984	468	-
1985	645	7
1986	615	1
1980-1986	2,424	12

The majority of written instructions and advice were issued from September through December, since most clearcut notifications are made during the autumn and winter. Several environmental features, for example flora habitats, are probably inadequately detected in the field during these months. (Spring and summer are the seasons to observe rich biotopes and rare plant species.) The uneven flow of clearcut notifications to the Forestry Agency makes inventories and environmental planning necessary, if protection of valuable habitats is to be attained. Most likely, shortcomings in this aspect contribute to the low number of environmental protection prescriptions and advice from the forest authority.

On an average, only one prescription was issued in every second county over the total period of six years, and fewer than ten written advice per Forestry Agency district. That averages anually to approximately one or two written advice documents per DFO. Even if non-documented oral advice to forest owners is added, the magnitude of these figures indicates that the implementation of the Forestry Act environmental requirements cannot generally have been significantly influenced by specific advice from the Forestry Agency. Steering forest-environmental protection through legislation is thus exercised through informal negotiations between the District Forest officers and the forest owners or their managers. The negotiation process is, to a large extent, determined by the forest owners' economic considerations rather than the total social benefits from the forest.

Forest environmental legislation in the United States

To help suggest and analyse causal relationships between different independent factors and policy outcome, it is often fruitful to make comparative analysis. Here, the comparison is made in a top-down perspective, comparing two countries which, to a certain extent, have selected different policy strategies for coping with the environmental degradation problem in forestry. It is interesting to analyse whether different policy strategies lead to different environmental outcome. In addition, useful lessons are often learned from anothers' successes and failures. Since it is difficult to compare the whole United States with Sweden, the comparison is made primarily with one state - the state of California.

During the comparison between the two countries, it must be held in mind that the political systems are quite different. In the United States, much of the policy is formed by legal appeals and public participation after the legislation is decided by government, whereas in Sweden eventual conflicts are presumably solved during the process of rule-making. Policy is essentially prospective and preventive in Sweden, whereas in the U.S. it is more retrospective and remedial. The governments' choice of implements also differ; in Sweden there is relatively greater reliance on informal negotiation to resolve implementation problems, and in the U.S. there is relatively greater reliance on changes in policy, formal rules and procedures (Elmore et al., 1986).

Moreover, the U.S. environmental legislation requires Environmental Impact Assessment (EIS) before projects and programmes may be implemented, including evaluation of environmental impacts and alternatives to the proposed action. The EIS procedures have become institutionalized during the past decade as a component of routine federal policy making in the U.S., and allow for public participation and inter-agency review of environmental impact statements (Culhane et al., 1985). In Sweden, EIS procedures are not formalized with regard to timing, comprehensiveness and participation. The remiss system, which would provide links between different agencies and the public, works only to the extent that it is initiated by government and its agencies themselves (Lundqvist 1979).

One reason for examining the forest environmental legislation in the United States is that the concept of multiple use forestry - taking into account both timber and non-timber resources - emanates from there; it was established as a guiding principle for the National Forests as early as in 1905:

> Where conflicting interests must be reconciled the question will always be decided from the standpoint of the greatest good of the greatest number in the long run (quoted from Cliff 1962, p.175)

In the U.S. Forest Service multiple use has been long practiced, compared to Swedish conditions (Compare Hultman 1984). Management takes into account, to a greater degree than is common in Sweden, social and recreational use of the forests. In addition, protection of a large number of other forest aspects are required; for example, old-growth forest, watersheds, fish habitat, wildlife and endangered species habitat, wilderness and cultural resources.

The National Environmental Policy Act (NEPA) of 1969 requires assessment of the environmental impacts of all federal activities, as well as public participation in the process. This implies, for example, that Land and Resource Management Plans and Environmental Impact Statements must be drafted, publicly reviewed, and politically accepted before forest activities on National Forests and Bureau of Land Management's (BLM) [4] forests may be effectuated. [5] Such plans comprise an assessment of alternative forest management strategies, ranging from priority given to environmental-protection on one hand and timber-production on the other. Alternatives are presented in the form that public participation is encouraged; they are further discussed through public hearings, other agencies' review and comments from environmentalists, forest industry and "ordinary" citizens.

As a result of the influence from public participation, amenity values are generally highly appreciated and strengthened during the review process. Federal land managers can use public involvement to produce more balanced decisions (Culhane 1981). Currently, the draft and review of Land and Resource Management Plans is under process for the majority of National Forests and the BLM forests, and the result yet remains to be settled. As noted by Haigh and Krutilla (1980), efficient resource management and public participation is, to some extent, contradictory within the government policy directives. The legislation governing the Forest Service activities could be characterized as a "consensus package that on analysis reveals many inconsistencies" (Ibid., p.409). Following the U.S. tradition, the courts will be final arbiters in disputes over the contradictory legislative content.

Private forest land is, however, not regulated by these federal forest laws. Instead, each state decides their own forest practices acts. There has long been a reluctance to regulate private forest practices in the U.S. against externalities or depletion of natural resources. The aversion is based on the constitutional private property right not to be taken for public use without just compensation. More recently, however, increased environmental awareness, the NEPA of 1969 and the 1972 Federal Water Pollution Control Act (Section 208) [6] have put pressure on the states to enact modern forest practice laws (Siegel and Cubbage 1984).

Starting with Oregon (1971), seven states [7] currently have amended comprehensive modern forest practice laws, including protection of both timber and non-timber resources and a rigorous structure for rule promulgation, administration and enforcement (Henly and Ellefson 1986). In addition, land zoning regulation controls timber harvesting and road construction activities on forest land in one state (Maine) and, locally, in many counties and municipalities throughout the U.S.(Siegel and Cubbage 1984).

In the older forest practices regulations, the state legislature specified all details directly in the law. Modern forest laws, however, provide for subsequent rule-making, or combine minimum standards specified in the state law with additional rule-making (Ellefson 1984). The rule-making authority is delegated to a forestry practice board, similar to Sweden's frame-law system. Depending on which state, these boards may receive technical advice from the state division of forestry, state or regional technical advisory boards, other state agencies, public hearings and written testimony, or from all of the above. The approach to administering the state forest practice law also varies between the states. Some of them require compliance with standards, others require notifications, and still others require a permit before operations may begin. In addition, different government agencies administer the law in the different states (Ellefson 1984).

Through questionnaires and interviews, Henly and Ellefson (1986) examined the administrative structure and effectiveness of forest practice laws in the seven states previously mentioned; they found almost universal agreement that the implementation of state forest practice regulation is responsible for significant improvements in the condition and protection of non-timber resources. These achievements were not generally accomplished through rigid enforcement, but rather through co-operative efforts on the part of the enforcing agency and the forest owners.

Nevertheless, most forest practice regulation programmes were found to be under funded. Also, the non-timber resource agencies generally have little more than advisory power in forest practice regulation. Consequently, legislative and administrative decisions of funding and power allocation between agencies all limit the extent to which both timber and non-timber resources can and will be protected through forest practice regulation (Henly and Ellefson 1986).

Comparison between California and Sweden

California is especially interesting because it has the most far-reaching forest environmental legislation in the U.S. (Henly and Ellefson 1986). The California Forest Practices Act was amended in 1973 as to environmental concerns in forestry. The size of California corresponds approximately to that of Sweden, and it is likewise dominated by forestry. Hence, there are general features which make a comparison worth while. The focus of analysis is to compare the factors which can explain different policy outcome in Sweden and California, following the framework used in the Swedish study (See Figure 1.2).

Based on an evaluative study of the influence that the California forest-environmental legislation has had on private forest lands (Green 1982), similarities and differences between California and Sweden are described and analysed as to: (1) form and content of the forest environmental legislation; (2) political and administrative culture and structure in forestry; and (3) pressure from environmentalists and from the public. Comparison is also made - although in a more limited way because of different research approach in the Swedish and California studies - between (4) harvest technology; (5) forest economic considerations; and (6) implementing actors' attitudes to environmental protection.

The strict regulation of forest practices in California may be related to the significant pressure from commercial exploitation of the land, from public demands for recreation as well as from environmentalists in the form of the Sierra Club and other conservation groups, who request greater environmental protection concerns. The act regulates silvicultural and related practices on non-federal property (both public and private) throughout the state by instituting planning procedures and by providing for the promulgation of rules by the California Board of Forestry.

The principal lumber-producing counties of the Redwood Region in California is the focus of Green's research. Based on fifty in-depth interviews [8] and sixteen telephone interviews with randomly selected professional foresters [9], he concludes that the act and the rules significantly improved environmental protection and altered forest management strategies. Much of the change is not caused by the rule requirements as such, but by the act's procedural requirements and an increased emphasis on environmental protection (Green 1982).

Before the legal requirements and the environmental achievements can be compared, general forest conditions of the Redwood Region must be contrasted to those of Sweden. The Redwood Region is recognized as a separate district in the California Board of Forestry's regulations, due to its ecological sensitivity. The total area of redwood forests has diminished drastically during the last century, which makes the remaining forests increasingly important to protect for environmental reasons. The conflict between amenity and commodity forest values in the Redwood Region is a cause for the environmentalists to react. For example, there is an organisation called the Redwood League, which concentrates on saving of redwood forests.

The influence of the stumpage value on environmental practices is discussed by Green (1982). As stumpage value increases, it tends to encourage forest practices that enhance timber production. At the same time, the increase in stumpage value leads timber operators to treat the timber, and the forest, with more respect compared to when the stumpage represents little economic value. Also, the enhanced value of the resource provides better economic profits, and hence more money is available to manage the land. Extra profits allow the forest owners to practice biologically sound forestry rather than focus on minimizing costs (Ibid., p.11). Compared to Sweden, the Redwood Region of California represents higher stumpage values in average. Possibly, the forest economic considerations therefore differ. On the other hand, it may also be argued that increased stumpage prices can make non-timber resources comparatively less valuable, and that the forest managers might be less willing to make sacrifices to protect environmental values (Ibid., p.13).

There are several similarities between the California and the Swedish Forestry Acts in terms of form and content of the forest environmental legislation; their wordings clearly place the value of timber production above that of other resources. In both acts, many rules with regard to environmental protection are specified by the rule-making Board of Forestry, and the content of the rules is quite similar. However, the California rules (to a greater extent than the Swedish) establish clear guidelines for environmental protection. For example, clearcuts are limited to 40 hectares [10] and stream protection requires at least 50 per cent of the overstory canopy shading the watercourse, as well as 50 per cent of the understory vegetation present before timber operations, to be left standing within the stream protection zone [11]. Those specified rules are written into the act as a result of review from other state agencies concerned with protection of wildlife, fish, water quality and erosion control. Inter-agency review has thus influenced the California Department of Forestry to adopt additional rules in order to protect non-timber resources (Ibid., p.31).

To compare the environmental protection achieved through the Swedish and California acts, it is also necessary to analyse differences in administrative culture and structure in forestry. The Swedish legislation and institutional arrangements are already described (See Chapter 3). In California, a proposed harvesting must not only be notified, as in Sweden, but it also requires a Timber Harvesting Plan (THP) prepared by a Registered Professional Officer (RPF) to be reviewed by the California Department of Forestry (CDF) before logging is permitted. Prior to the 1973 act, no such pre-planning of the operation was required. Any delay in harvest operations can be costly, so there is a strong incentive to prepare high-quality THPs for approval by the CDF. The THP forces foresters to consider environmental aspects; for instance, it requires information on trees used for nests by eagles, osprey, or endangered bird species, as well as data on streams and erosion hazard. Through the planning, foresters anticipate problems before they occur (Green 1982).

The concept of the Registered Professional Forester was created by a state act, which became effective at the same time as the Forest Practices Act in 1973. RPFs are licenced by a "foresters licencing committee" at the California Board of Forestry, consisting of members both from forestry and from the public. Licence is given after seven years of professional work, or after three years if the person has a forest degree, which enables the state authorities to prevent non-complying foresters to become licenced. [12]

Preparing the six-page THP is estimated to take around twenty man-days. This may be compared to the Swedish clearcut notification, which requires only one page. In addition to the CDF officer, a review team evaluates every THP, bringing in additional expertise from the Departments of Water Quality and Fish and Game. These experts include biologists, geologists, archeologists, hydrologists, engineers and soil scientists. The function of the review team is to evaluate the environmental impacts of timber operations proposed in THPs. Nevertheless, the actual power of these other agencies is quite minimal. The strongest action they can take within the review team is to write a non-concurrance, which has no formal effect. The most important function these other agencies can perform is to suggest mitigation measures to avoid environmental damage During the review process, many THPs are modified by the Department of Forestry before approval. [13]

Monitoring compliance with the regulation also differs between Sweden and California. In Sweden, the Forestry Agency officers are not forced by procedural requirements to inspect the harvestings. Field inspections are carried out on the officer's own initiative. [14] In California, the act provides for a minimum of three periodic inspections of logging operations and an optional pre-harvest inspection by the CDF. Since 1973, when the act was instituted, the inspection force has increased sixfold. In 1985 approximately 80 per cent of the logging operations were inspected before harvest. [15] Representatives from Water Quality and Fish and Game are able to attend only about 10 per cent of the pre-harvest inspections, because of limited staffing. The pre-harvest inspections serve as a useful function, as they avoid friction and potential conflict with state agencies and the general public (Green 1982, pp.18-9).

Participation by the public is another area where the forest regulation differs between Sweden and California. In Sweden, no public announcement of proposed clearcuttings is required, other than the clearcut notifications submitted to the County Forestry Agency be made official, so that anybody is permitted to personally check the filed documents. Also, there are no formal requirements for the Forestry Agency, or the forest owners, to gather comments from the public, environmentalists or neighbours, who may perceive themselves affected by the cutting. In California, notice of the THP filing must be posted in the CDF offices, and comments by the public must be considered according to the act. The CDF is also required to send notices of proposed THPs to all interested parties [16] and to make THPs available for public review. In addition, recent rule changes have increased the responsibility of the timber landowner to notify neighbouring landowners of proposed logging operations. (People who live in the 100 meter vicinity of the proposed cutting are provided with a copy of the THP.) At all roads leading out of the area, a short version of the THP is posted. As a result, public involvement has increased, and potential conflicts are usually being solved at an early stage. Participation in the review process by agencies other than the CDF, and by the public, make the RPFs anticipate the concerns of others. Consequently, forest practices reflect non-timber values in forestry to a greater extent today, compared to before these procedures were initiated.

In Sweden, the forest owner is the only person who can be legally controlled by the Forestry Agency through a penalty. [17] In California, however, the people who conduct the harvesting can also be punished by the forest authority. Every timber operator who carries out the harvesting is responsible to follow the Timber Harvesting Plan. Moreover, he must have a licence from the CDF in order to be allowed to harvest. Timber operators, who deliberately neglect environmental protection can thus be stopped from working, if their licence is suspended.

In California, as in Sweden, legal punishment is rarely used to make the forest owners comply with the environmental regulations. Very few clearcuttings are totally prohibited and few cases are taken to court trial. [18] RPF and timber operator licences are seldom suspended because of environmental violations. However, the reason why so few clearcuttings are prohibited or taken to court differs between California and Sweden. In Sweden, it is not effected because the forest and administrative culture within the Forestry Agency prevents the officers from writing formal prescriptions or prohibitions on environmental matters. Also, "outside" interests (such as environmentalists) are not legally permitted to appeal against a timber harvesting in case of environmental damage. In California, prohibitions and law-suits are rare due to the act's rigorous procedural requirements with regard to environmental protection, which emphasize prevention before the fact rather than after the damage is done. Also, in California, inter-agency review and public participation promote the concerns for non-timber values in forestry, and put pressure on the forest owners, foresters and timber operators to consider environmental aspects. According to the interviews, the threat of rule enforcement has no significant impact on improving forest practices, apart from eliminating gross violations and forcing timber operators with bad records out of the business (Green 1982, p.14).

The conclusion of the California study is that rigid rules and regulations are not crucial to the achievement of policy goals; on the contrary, they may even be perceived excessive by the forestry practitioners as a sign of overregulation. The results indicate that the goals of environmental protection may be equally well realized through flexible procedural guidelines (Ibid., pp.30 and 40). This conclusion is quite important, since forestry is not enabled to dictate all practices - there is too much site variability. The potential for procedural requirements to effect real improvements in forest practices is, however, significant (Ibid., p.40).

Like the Swedish study, the interviews in California indicate that implementing actor's attitudes and forest economic considerations are significant factors, which contribute to the environmental protection behaviour. A major change of attitudes towards environmental protection has occurred among the California foresters:

> Foresters no longer simply look at the lands in terms of potential harvest. They have always claimed to be practicing ecologists but in the end, according to several foresters and most state agency personnel interviewed, they were interested only in volume harvested. Virtually all informants argued that foresters are now more sensitive to non-timber concerns such as wildlife and watershed than they previously had been... One forester cited the treatment of dead snags as an example of his shift. In the past they were always removed because of fire hazard. Now they are often left to provide wildlife habitat (Ibid., p.16).

In the California study, the influence of different harvest technology was not evaluated. Partly, this may be because logging in the Redwood Region is performed manually - the steep slopes and erosion hazard do not allow mechanized cutting. The location and design of main haul roads in this region is evaluated in the THP; consequently, environmental damage from forest machines is taken into account. Forest economic considerations, however, are viewed as a limiting factor in California, especially among the small landowners. Overall costs due to the act and the rules are estimated to a 25-33 per cent increase of administrative and logging costs, which represents how much landowners with a permanent interest in timber production would be paying for a typical harvest operation that includes some sensitive areas (Ibid., p.24).

The California study suggests that the state authorities should investigate ways in which the costs to small landowners of compliance with forest practice procedures could be reduced, and to institute efforts to reduce the cost of procedures to landowners with good forest practice records (Ibid., p.41). The similarity with Swedish conditions is obvious. In Sweden, small landowners are sometimes severely economically affected by requirements for environmental protection, but cannot be compensated by the State. In addition, competing goals of high and valuable timber production in the Swedish Forestry Act may even require them to harvest their forest.

Conclusions

The conclusion from the above analysis is that the environmental legislation is used as a basis for negotiations in both California and Sweden, with very few legal sanctions. Nevertheless, there are substantial differences between the ways in which the rules are used. One aspect that differs is <u>who</u> exercises the control of environmental behaviour-whether the steering is primarily channelled through the bureaucrats and the experts, or if steering also takes place during the process by the people whose environment is affected by forest operations.

Another aspect is to consider <u>to what extent</u> the environmental protection requirements imply economic sacrifice (compared to the forest management which otherwise would have been practiced). Environmental requirements either impose major economic sacrifice from the target groups or they may be essentially minor and not economically demanding. A third aspect is to what extent the forest managers perceive a "real threat" of legal punishment. Although the use of sanctions is similarly low between California and Sweden, the potential for sanctions differs, and may consequently affect actual behaviour.

In California the forest-environmental regulation is effectuated both by hierarchical control through the bureaucracy, and by public participation and environmental group pressure throughout the policy formulation and implementation process. The regulation requires a substantial economic input, compared to Sweden, because of the thorough environmental assessment and pre-harvest planning requirements. Also, the threat of the THP being denied by the California Department of Forestry after the review process forces the foresters to consider environmental aspects in the plan. Moreover, both the public and environmental groups are allowed to appeal, implying additional potential threats of sanction to the forest managers.

In Sweden, the legislation is almost exclusively used without the threat of sanctions. Monitoring compliance is less extensive than in California, and primarily exercised through information activities by the forest and environmental authorities. The public and the environmentalists are supposed to influence the policy and legislation ex ante through Royal Commissions and through the remiss process when the legislation is formed. Thereafter, their interests are presumably channelled through the agency who is to implement the programme. All aspects of the public interest are thus supposed to be already taken into account in the decisions made by government and its agencies. Hence, participation by the public in the implementation process is not considered necessary.

The same Swedish strategy to deal with the public interests is found by Lundqvist (1980) in his comparison of clean air policies in Sweden and the U.S. (pp.85-92). Furthermore, Ericsson (1985) concludes in his investigation of Swedish environmental policy that although a corporation which applies for a licence to pollute the environment has to discuss the problems in public before the application is made [19], public interests are in reality only represented by the different government agencies. Even the environmental groups claim protection of Nature by essentially conforming to the formally established democratic process; thereby, they do not put a real pressure on the government and its agencies - they are themselves found within the "system" (Ericsson 1985). Public participation in Sweden is restricted to indirect influence through the political parties and interest groups, and the agencies are neither required to organize public hearings nor to obtain information and suggestions from the public during the implementation process.

The Swedish legislation clearly limits the environmental measures within the Forestry Act requirements to "not considerably damage economic forestry". By definition, the forest owners are not supposed to undertake protective measures which considerably restrict their forest management. Pre-harvest investigation and marking of environmentally valuable areas is not formally required. The Forestry Agency relies on informal negotiations and advice to forest owners and others who are engaged in forest operations.

However, there are tendencies also in Sweden to incorporate environmental aspects into forest management planning and to decentralize political decision-making. For example, the recently amended Building Act (1987) allows the municipalities to influence forest management practices within the municipality boundary through land use planning in consultation with affected parties (such as the County Administration, the County Forestry Agency, forest owners and municipality citizens). [20] The result of this planning process remains to be seen, but it may - possibly - imply more participation by others than the forest and environ- mental agencies in decisions over forest practices that affect environmental values.

In California the control of environmental protection is exercised not only by the forest authority, but also by non-timber agencies and experts on different aspects of the environment. Concurrent government and reciprocal review as it is practiced in the United States may reduce some of the problems of elitist representation and error oversight. The policentric network also has closer contacts with those affected and is able to judge factual considerations from several perspectives (Boschken 1982). Contrary to what Pressman and Wildavsky (1973) suggest - that the more actors involved and the more complexity of the implementation process decrease chances for action - under certain conditions in regulatory policy complexity may actually ease implementation by rendering multi-actor compliance more attractive (Durant 1984; O'Toole and Montjoy 1984).

The vague content of the regulation and the soft steering strategy of the forest authorities emphasize the influence of time as a factor which may be important. In California the forest-environmental legislation is a couple of years older than it is in Sweden and therefore, forest professionals' attitudes have been influenced by the legislation for a longer period of time. Nevertheless, attitude changes are likely to occur faster if incentive structures favour priority to environmental protection over timber production (in environmentally valuable areas), as is the case in California. At present, no such major incentives are discerned in Sweden; hence, attitude changes among the Swedish forest professionals are likely to wait in developing.

Conditions for success with the different regulatory strategies in the two countries investigated depend, in addition to formal rules and requirements, on a number of other factors which are mentioned in the beginning of the chapter: the capacity of the forest authority which is to control the environmental performance; the qualifications of the people in charge of the forest operations and their attitudes; socio-economic and technical conditions; and pressure from the public and from environmentalists. Hence, steering environmental protection through legislation and rule-making is likely to be successful only if these other factors work together and strengthen the environmental goals during the implementation process. Similarly, it is likely to fail if agency resources, attitudes of implementing actors, forest economic considerations and technology counteract and restrain environmental protection.

NOTES

[1] The percentage refers to the share of the specified environmental measure in proportion to the total number of environmental measures required in the written instructions and advice (2,329 measures).

[2] The following five prescriptions contained recreational aspects: one which demanded retention of seed trees on a clearcut along a heavily travelled road; three which required clearing of logging debris; restoration of

deep tracks and retention of aesthetic trees; and finally, one which required a smaller area clearcut than applied for in the notification because the area was adjacent to a densely populated area.

[3] The biologically-motivated prescriptions included the following cases: protection of small forested islands, a bog-dominated forest area, single trees on a forest area which was to be transformed into a pasture, a goshawk nest on an island, a capercaille courtship area, a floristically rich bog adjacent to a lake, and prohibition of ploughing after clearfelling in an area which also was used for reindeer husbandry.

[4] The Bureau of Land Management is a federal agency, which among other public domain lands manages forest land, especially in the western states.

[5] Naturally, forest activities are allowed also during the time when a new plan is being drafted and reviewed; however, such activities may not be carried out on areas with great conflicts until the "best" management strategy is decided.

[6] This act required states to establish programmes for the control of nonpoint source pollutants arising from silvicultural operations.

[7] The other six states are Idaho (1974), Nevada (1971), Alaska (1978), California (1973), Massachusetts (1983) and Washington (1974).

[8] The in-depth interviews were made with registered professional foresters, state employees responsible for the implementation of the act, lobbyists and academic foresters.

[9] The professional foresters work as consultants to small-forest landowners. The sample represented 44 per cent of the full-time consulting foresters in the Coast District, which was the forest district that was examined.

[10] In sensitive areas, smaller size clearcuts are required.

[11] The stream protection zone shall be defined, according to the rules, by a Registered Professional Forester, who has to clearly identify the zone for each clear-cutting on the ground prior to the start of timber operations.

[12] Facts on the RPF licence originate from a pamphlet: "The Professional Forester in California". California Board of Forestry.

[13] In 1985, for example, 55 per cent of the approved THPs were changed by additional information (California Forest Practice Program Report, The Resources Agency and Department of Forestry, State of California, 1985).

[14] Of course, there are general recommendations from superior levels of the organisation that inspections shall be carried out. The ultimate decision is, however, made by the lower-level officers from case to case.

[15] California Forest Practice Program Report, 1985.

[16] This is effectuated by a mailing list, to which anyone can join.

[17] As earlier described (See Chapter 3), he can only be fined if he does not comply with a prescription or prohibition, issued by the Forestry Agency.

[18] In 1985, for example, only three Timber Harvesting Plans out of 1,234, that were submitted to the CDF for approval were denied; also, not more than four lawsuits aimed at stopping timber operations and the processing of timer harvesting plans were active (California Forest Practice Program, 1985).

[19] According to the Environmental Protection Act (§ 12a).

[20] Before the formal amendment of the Building Act there were very similar informal activities undertaken when municipal land use planning was carried out.

8 Summary

Public concern for the preservation of environmental values in forest operations has become a major political issue during the last decades. National parks and nature reserves are created for nature conservation, but these areas do not suffice - the remaining forest land must also be managed with care for environmental values. The government has answered by amending environmental protection requirements in the Swedish Forestry Act of 1974. A new paragraph in 1979 - § 21 - along with a penalty clause, stipulates that environmental considerations must be shown with all forest operations regardless of forest ownership and without economic compensation.

The aim of this thesis is to investigate *to what extent* and *why* environmental protection is (or is not) achieved in clearcuttings. The study describes and analyses the implementation of environmental protection in Swedish forestry through examining what actors are involved and what factors contribute to environmental protection outcome. Monitoring environmental protection in forestry can take place not only through legislative requirements and rule-making by the environmental and forest authorities; it may as well be exerted by control and/or influence on technology, economic conditions, attitudes, education and administrative culture of the implementing organisations and individuals. Moreover, steering can be effected both from the top and down, through the bureaucrats and the experts, and from the bottom and up during the implementation process, by people whose environment is affected by forest operations.

Different approaches to the analysis of policy and implementation reflect divergent basic stands towards the possibilities for the modern welfare state to identify and solve problems. In a top-down perspective, "policy" is the general term for collective goals, defined by public institutions. It may include one or several "programmes" launched by government. Studying the "implementation" process implies investigating what happens after the policy or the programme is set up. "Policy output" refers to the ultimate policy consequences of a particular policy or programme. "Policy outcome" includes, in addition, unintended results of other policies or programmes which may impede or adjust the actual forest-environmental policy outcome. Depending on the focus of analysis, such factors may be more or less difficult to discover.

The top-down approach assumes an ideal type of world according to which democratic political decisions will gradually be specified, implemented (at least approximately) and enforced by public authorities; it concentrates on attainment of formal objectives and administrative control why informal factors and unintended results may easily be overlooked.

The bottom-up approach, on the other hand, views the implementing actors at the end of the policy "chain" as primary actors whose knowledge, incentive structures and bargaining relationships become crucial factors when environmental considerations are evaluated. The bottom-up method begins with identifying these actors and puts more stress on informal rules than on programmes and formal decisions.

Policy, in the bottom-up perspective, may well exist without any programme being initiated. Collective problems are not necessarily defined by formal decisions. Analysis of the implementation process implies that collective actions are studies at policy delivery level. Implementation structures are defined by inductive research, through examining which actors organise to solve a particular problem or issue.

In my study of the implementation of environmental concerns in Swedish forestry, I make use of both the top-down and the bottom-up approach. I evaluate the extent of environmental considerations based on the forest-environmental legislation and prescriptions issued by government authorities, combined with preservation recommendations made by researchers. Even though the policy problem is defined by formal statements, the methodology is strongly inspired by the bottom-up approach. I start with examining environmental protection performance in the field, and identify crucial factors contributing to the result by interviewing the people involved at the policy delivery point.

My framework for analysis includes factors which directly, or indirectly, may influence environmental outcome in clearcutting operations, which is the dependent variable. Environmental protection performance is assumed to be explained by four independent variables: (1) harvest technology; (2) forest economic considerations; (3) implementing actors' attitudes and knowledge; and (4) influence from environmentalists. The societal frame, legislation and forestry culture and structure are regarded as underlying independent variables which may indirectly affect all these independent variables.

The study combines field investigation of clearcuts, interviews with implementing actors and evaluation of Forestry Agency documents on environmental protection issues. Interviews were carried out both through network analysis starting from case studies of clearcuttings and with different forest and environmental organisations. The network interviews were made with all those responsible for the outcome of the specific cutting, both from the County Forestry Agency, which is the government authority monitoring the Forestry Act, and from public and private forest owners, managers and operators.

Environmental protection results

Using a stratified random sample of 158 clearcuttings, distributed among five counties [1] and different forest owner categories [2], environmental protection was evaluated both before and after harvest. The results show that roughly half of the environmental features required were protected, according to Forestry Act regulations and preservation recommendations made by researchers. As a whole, approximately two out of five clearcuttings did not comply with the requirements.

Flora and fauna habitats (for example buffer zones along water courses and nesting trees for birds) were the least protected compared to aestetic values (such as single, beautiful trees and buffers adjacent to lakes and residential areas). Bogs, fens and bedrock areas, which are most often technically and/or economically unsound to harvest, were frequently retained in accordance with the environmental regulation.

The explanatory power of different factors to the result of environmental protection within the 158 clearfellings was analysed by a probit model, using multiple regression. This analysis was limited to factors possible to determine at the time of field inspection, without any interviews. Mechanized harvest operations with harvesters and processors were found to affect environmental protection negatively compared to manual felling and processing with chain saws. Nesting trees for birds were rarely retained with these mechanized harvestings; logging residues were rarely cleared away from watercourses and forest paths; and deep tracks and ground damage were prevalent. Differences in environmental protection achieved between counties and forest owner categories (including which category had carried out the harvest) were mainly explained by the degree of mechanization. The greater the proportion of manual cuttings, the more environmental protection was considered. Nevertheless, the results also indicate slight differences in forest-environmental culture and attitudes between the counties and forest owners in terms of care for flora and fauna, which are the environmental aspects that probably require the most ecological knowledge to protect.

Implementing actors' attitudes

These results were followed-up by interviews related to twenty-five of the investigated clearcuttings. These case studies illustrate the natural divergence in environmental conditions and suggest multiple factors, which - to various extent - can explain the result of environmental protection. A large number of actors are involved in the field level implementation of environmental protection in forestry. These actors include district officers at the County Forestry Agency, forest owners, forest managers and administrators (such as foresters, planners and timber officers), and forest workers. Occasionally, officers from other public agencies, environmentalists and other private actors are also involved.

A "key actor" was identified whose decisions substantially affected the behaviour of other implementing actors. The key actor's attitudes to environmental protection, his forest economic considerations and his contacts with environmentalists and environmental authorities were specifically examined for each of the twenty-five case studies.

Three case studies were selected, ranging from "failure" to "success", and presented in detail as to what happened and why during the implementation process. These three clearcuttings were all performed with a harvester and showed that mechanized harvesting does not necessarily imply environmental degradation if only the environmental measures are well planned for. Generally, no single factor was found to be the only determinant to the result of environmental protection. Rather, a variety of different factors worked together, or counteracted during the implementation process.

The implementing actors' attitudes to environmental protection were examined. Their perceptions of the importance of different environmental measures generally corresponded to what they actually protected in reality - except the regard for nesting trees. Protection of nesting trees was highly valued by many of the respondents but rarely observed in the field, due to difficulties in detecting and retaining them with the large forest machines. There were no differences in attitudes between the manual forest workers (working with chain saws) and those who operated forest machines. This indicates that the difference in environmental protection achieved with manual and mechanized harvestings is not, to any large extent, a matter of workers' attitudes but rather caused by the harvesting method as such.

More than half of the respondents had not received any basic education in environmental protection. Those who were most informed on environmental protection were the Forestry Agency officers, followed by the supervising forest managers; they valued a greater number of environmental aspects than those who were less informed. The forest workers were generally less informed and educated in environmental protection, as were also the private, individual forest owners. There were no significant differences between young and old forest professionals; thus, the attitudes to environmental protection will not immediately change as an alteration of generations takes place in forestry.

The analysis also indicates that even the ones who were relatively better educated in environmental protection perceived that they had neither enough knowledge, nor time to pay attention to environmental values. Many of the most educated forest administrators and managers experienced their lack of staff as a major problem. Lack of adequate environmental planning and divergent attitudes to environmental protection among the different forest professionals involved in harvest operations thus contributed to environmental protection failures.

Environmental organisations do not (according to the interviews) generally pressure forest professionals to protect environmental features. The majority of respondents reported none, or only a few specific contacts with environmentalists. Frequent consultations were very rare. Almost two-thirds of the persons interviewed were hunters, but few were members of environmental organisations other than hunting.

Indirect influence on environmental protection performance from environmentalists was, however, mentioned by some of the respondents. For example, environmentalists contribute to policy formation and implementation both through political pressure to legislate and through writing and/or initiating mass-media reports which affect the attitudes of forest professionals.

The respondents frequently cited harvest technology and forest economic considerations as important factors affecting environmental protection. Manual harvesting with chain saws was generally regarded to be superior to highly mechanized harvest operation systems. The forest machines were considered inadequate in several ways: single, environmentally valuable trees are difficult to detect and to retain and the machines cause deep tracks. In addition, demand for high economic profit puts pressure on the machine operators to work at a fast pace. Many of the forest workers are paid piece-rate and the machines operate all year round, often at dark and regardless of weather. Also, sufficient pre-harvest marking of environmentally valuable trees and areas to facilitate protection is not carried out.

Economic conflicts were perceived both in the smaller-scale of retaining single trees and in larger-scale, such as protecting several hectares of forest land. The respondents' perceptions on how much could be protected for environmental reasons without economic compensation were generally quite moderate; for example, a few hectares at the most. The lack of resources for compensating economic losses when several hectares should be protected is a considerable problem, especially for private, individual forest owners, generally owning small holdings. Many of the environmentally most valuable forest areas cannot be protected since there are no state fundings for preserving such small areas. Forest taxation currently also counteracts environmental protection, since forest owners are taxed in proportion to the commercial value of their forest holdings, regardless of whether the forest should be (or already is) protected for environmental reasons.

Steering through legislation

Since the Forestry Act environmental regulation is designed as a "frame-law" with ambiguous goals, it has to be further interpreted by the implementing agencies. The National Board of Forestry is the rule-making authority and its County Forestry Agencies with officers at district level transform the central directives into practical measures, such as advice and information to forest owners. In addition, the Environmental Protection Section at the County Administration is involved with environmental protection in forestry according to the Nature Conservancy Act, by which nature reserves and national parks may be created with economic compensation from the State.

Proposed clearcuttings must be notified in advance to the County Forestry Agency, which thereby may influence environmental protection measures. Forestry Agency officers at regional and district level are responsible for the daily supervision of the Forestry Act. Since their work is characterized by increasing demand for services, inadequate resources in comparison to their tasks and conflicting goal expectations, they have to make priorities.

Environmental protection most often is perceived as a constraint for timber production, and therefore often given low priority compared to other forest management goals. The Forestry Agency officers are also torn between the role of "policeman" and the role of advisor. Their organisational culture favours the role of the friendly counsellor and the influence on forest owners is characterized by soft steering methods. Prescriptions, which if neglected may lead to a fine, are very rarely issued. Moreover, no cases have been taken to court. Instructions and advice are more common but cannot be legally enforced. Such advice may be either oral or written.

All twelve prescriptions and 842 advice documents issued from 1980 through 1986 (representing 35 per cent of all written advice during the period) were investigated. The content of these documents reflects the variety of environmental measures required by the Forestry Act § 21. However, specific floristic values and protection of cultural resources were rarely mentioned. The amount of written environmental advice varied significantly between different County Forestry Agencies and districts. The older District Forest officers had generally issued more written environmental advice on an average than the younger.

The frequency of written prescriptions and advice has gradually increased over the six years since the Forestry Act § 21 was amended. The content of environmental measures required remains approximately the same, however, which indicates no change of attitudes among Forestry Agency personnel in terms of what aspects of environmental protection are emphasized. On an average, one or two written environmental advice per District Forest officer were issued every year. Even if non-documented oral advice to forest owners is added, the magnitude of these figures indicates that the implementation of the environmental requirements cannot generally have been significantly influenced by specific advice from the Forestry Agency.

The extent of environmental measures required by these formal documents did not impose any considerable economic sacrifice to the forest owner. Steering forest-environmental protection through legislation is characterized by informal negotiations between the Forestry Agency officers and the forest owners. The negotiation process is, to a large extent, determined by the forest owners' private economic considerations rather than the total social benefits from the forest. The Forestry Agency officers' attitudes to environmental protection generally correspond to the forest managers' (foresters and timber officers in charge of forest planning and operations) according to the interviews. Hence, the forest owners do not perceive any significant pressure from the forest authority to protect environmental features.

Comparison between California and Sweden

A comparison between forest-environmental regulation in the United States and Sweden reveals both similarities and differences. [3] Multiple use is practiced on U.S. federally-owned forest land for almost one century. Modern, comprehensive forest practice laws are currently enacted in seven states; in addition, private forest land is regulated in terms of environmental aspects in many other states as well as counties and municipalities. The implementation of these state forest practice regulations are responsible for significant improvements in the conditions and protection of non-timber resources. These achievements are accomplished through cooperative efforts on the part of the enforcing agency and the forest owners, rather than rigid enforcement. Public participation is also favoured by the U.S. legislation and contributes to strengthening environmental protection requirements.

In both California and Sweden the forest authorities use soft steering methods to influence the forest professionals' environmental behaviour. Comparatively more formal requirements and greater threat of sanctions are, however, present in California. For example, anybody may appeal against the proposed harvesting. In Sweden, as a contrast, the environmental protection decisions are characterized by informal negotiations between the forest authority and forest managers. In Sweden less direct public participation is achieved and the pressure from environmentalists throughout the implementation process is less significant.

In the state of California, much of the change is caused by procedural requirements. Environmental impacts of proposed clearcuttings must be described and evaluated in Timber Harvesting Plans, prepared by registered foresters, and approved by the California Department of Forestry prior to the commencement of logging. In addition, expertise from non-forest agencies review the plan and may suggest mitigation measures to avoid environmental damage. The proposed clearcutting must be publicly announced, and citizens are encouraged to communicate their reactions. As a result, potential conflicts are usually solved before the cutting is allowed. Forest professionals in California are forced to consider environmental aspects, and their attitudes to environmental protection have changed significantly.

An important conclusion from the California study is that rigid rules and regulations were not found to be crucial to the achievement of policy goals; rather, the goals of environmental protection may be equally well realized through flexible procedural guidelines. However, the California study also indicates that significant environmental protection cannot be achieved without economic input, such as expenses for pre-harvest planning and review.

Conclusions

There is general political consensus in Sweden on that threatened plant and animal species should be preserved and a diversified forest ecosystem should be maintained. However, the means by which this is to be achieved and who is going to pay for it - the State or the forest owners - remain in dispute. Current forest-environmental legislation leaves room for discretionary judgements by lower-level officers and - according to the Swedish tradition - courts are not used for solving remaining conflicts.

According to the official Swedish government tradition potential conflicts are presumably solved during policy formation when different interests are heard, for example through the Swedish "remiss" system. Also, the government agencies are supposed to take public interest into consideration in additional rule-making and implementation of policy directives. In reality, this study indicated that "ordinary" citizens and environmentalists do not significantly impact the environmental design of Swedish clearcuts in general. Instead, forest owners' economic considerations and harvest technology were found to be the most important factors affecting policy outcome.

Economic profitability - expressed as a high and sustainable timber production - is the principle goal in Swedish forestry, and the conflict with environmental protection and nature preservation is therefore always present. However, as shown in this study, much environmental protection may be achieved without considerable negative effects to economic forestry. Highly mechanized harvest operations generally, but <u>not</u> always, contribute to environmental degradation; good knowledge of environmental values and careful planning sometimes help to protect environmental features.

The large number of actors involved in the implementation process requires a great deal of co-ordination. One or two persons' behaviour may destroy what others try to protect. Then, the attitudes and knowledge of <u>all</u> the forest professionals who supervise, plan and carry out forest operations are important. Their attitudes are changeable and, in this perspective, the forest-environmental legislation may act as an authorization to consider environmental aspects and to learn more.

Ultimately, the degree of environmental protection achieved depends largely on the extent of political pressure from the public to change forest management practices, both through economic incentives, design of technology and forest professionals' attitudes. The result of the forest-environmental policy thus depends on a number of factors which are not easily controlled from the top and down. Steering environmental protection through legislation and rule-making is likely to be successful only if there is a pressure to influence these other factors to work together and strengthen the environmental goals during the implementation process.

NOTES

[1] The five counties are Norrbotten, Västerbotten, Västernorrland, Värmland and Östergötland.

[2] Forest owner categories included private individuals, private companies, state-owned and other public forests.

[3] The evaluation of U.S. forest-environmental regulation is based primarily on Green 1982, Ellefson 1984, Siegel and Cubbage 1984 and Henly and Ellefson 1986.

References

I. UNPUBLISHED SOURCES (all in Swedish)

Department of Forest Economics, Swedish University of Agricultural Sciences, Umeå

Minutes from four forest and environmental authority conferences concerning "Environmental Protection in Forestry": Ljusdal, June 3-4, 1981; Bollnäs, 30 September-1 October, 1981; Linköping 7-8 October, 1981; Jönköping, 14-15 October, 1981.

Minutes from a conference on "Fauna Protection in Forestry", arranged by the Ornithological Association of Västerbotten and Norrbotten, Åsele, 4-5 April, 1981.

Results from 19 preparatory interviews, carried out in 1981 with forest and environmental authorities and forest managers (transcribed).

Interviews carried out during 1983-1985 with 24 Forestry Agency officers, 10 private, individual forest owners, 36 forest managers and 45 forest workers (tape-recorded and transcribed as well as computerized).

Field investigation of 216 proposed clearcuttings during 1982-1985, of which 158 were harvested (documented and described with photographs as well as computerized).

Review of original judgements (13 of the investigated clearcuttings) by forest and environmental authorities, August 1984.

Information from 842 Instructions and Advice on environmental protection issued by the Forestry Agencies during 1980-1986, computerized.

Selected documents, minutes and statistical data from the National Board of Forestry as well as from the County Forestry Agencies concerning environmental protection in forestry during 1981-1987.

Selected documents from the National Board of Antiquites concerning environmental protection in forestry during 1981-1987.

Selected documents from the Swedish Forest Workers' Labour Union concerning environmental protection in forestry during 1981-1987.

Selected documents from private forest companies concerning environmental protection in forestry during 1981-1987.

National Board of Forestry, Jönköping

Prescriptions and Instructions and Advice on environmental protection in forestry, issued by the regional County Forestry Agencies during 1981-1986.

Selected verdicts from court appeals, appeals to the Ministry of Agriculture and to the National Board of Forestry concerning environmental protection in forestry.

National Environmental Protection Board, Stockholm

Internal documents from the SKONA-group: Viewpoints from the forestry representatives, March 1977 and Viewpoints from the environmental representatives 1977-03-29; supplement 1977-09-12.

Minutes on protected national parks, nature reserves and other forest reserves, 1987.

II. PUBLISHED SOURCES

II A. Official Documents and Statistics (in Swedish)

Debate in Parliament

RD 1974:145, 1978/79:154.

Debate in the Standing Committee on Agriculture

JoU 1974:52; 1978/79:30.

Government Bills

Prop 1971:122; 1974:166; 1978/79:110.

Legislation

Building Act SFS 1987:10.

Environmental Protection Act SFS 1969:387.

Forestry Act SFS 1948:237, supplemented in 1974:1026.

Forestry Act SFS 1979:429 with National Board of Forestry regulations SKSFS 1983:2.

Nature Conservancy Act SFS 1964:822.

Parliament Bills

Motion 1974: 2003, 2005, 2008, 2010, 2011.

Motion 1978/79: 1438, 1458, 1964, 2262, 2274, 2281.

Royal Commissions

Skog för framtid (Forest for Future, SOU 1978:6-7. Liber/Allmänna förlaget, Stockholm.

Politisk styrning -administrativ självständighet (Political Steering - Administrative Autonomy), Betänkande av förvaltningsutredningen, SOU 1983:39, Liber/Allmänna förlaget, Stockholm.

Swedish Government Official Reports

Kalhyggen (Clear-fellings), Ministry of Agriculture, Ds Jo 1974:2.

Skogsbruket, LU 80 (Forestry in the Long-Term Planning Commission 1980), Ds E 1980:4, Supplement I, Liber/Allmänna förlaget, Stockholm.

Swedish Official Statistics

Statistical Reports: SOS J 1979:14.

Statistical Yearbook of Forestry, National Board of Forestry, Jönköping, 1987.

Other Reports

California Forest Practice Program Report, California Resources Agency and Department of Forestry, Sacramento, 1985.

Selected documents from the California Department of Forestry, 1987.

Naturvårdshänsyn inom skogsbruket - samarbetet mellan myndigheterna (Environmental protection in forestry - cooperation between the authorities), National Environmental Protection Board together with the National Board of Forestry, SNV pm 1827, Stockholm.

Natur- och landskapsvård (Nature Conservation and Landscape Protection), National Board of Forestry, 1974, Jönköping.

Skogsvårdsorganisationen 1985-1986 (The Forestry Agency Organisation 1985-1986), National Board of Forestry, 1987, Jönköping.

II B. Books and Articles

Ahlén, I. (1977), Faunavård (Fauna Protection), Liber, Stockholm. (in Swedish).

Ahlén, I., Boström, U., Ehnström, B. and Pettersson, B. (1979), Faunavård i skogsbruket (Fauna Protection in Forestry), National Board of Forestry, Jönköping. (in Swedish).

Andersson, S. (1982), Ny teknik i skogen (New Technology in the Forest), Report 702 from project "Samhället och skogen" (Society and Forest), Secretariat for Future Research, Liber, Stockholm. (in Swedish).

Andrews, R.N.L. (1976), Environmental Policy and Administrative Change, Lexington Books, Lexington, Mass.

Anton, T.J. (1980), Administered Politics: Elite Political Culture in Sweden, Martinus Nijhoff, Boston, Mass.

Bardach, E. (1977), The Implementation Game, MIT Press, Cambridge, Mass.

Boschken, H.L. (1982), Land Use Conflicts: Organizational Design and Resource Management, Univ. of Illinois Press, Urbana.

Cliff, P. (1962), "Multiple-Use Management on the National Forests of the United States", in Fifth World Forestry Congress, Proc:s, Seattle, Wash. 1960. pp.173-81.

Culhane, P.J. (1981), Public Lands Politics: Interest Group Influence on the Forest Service and the Bureau of Land Management, Resources for the Future, Johns Hopkins Univ Press, Baltimore and London.

Culhane, P.J., Armentano, T.V. and Friesema, H.P. (1985), "State-of-the-art Science and Environmental Assessments: the Case of Acid Deposition", Environmental Management 9:5 pp.365-78.

Durant, R. (1984), "EPA, TVA and Pollution Control: Implications for a Theory of Regulatory Policy Implementation", Public Administration Review 44:4, pp.305-15.

Eckerberg, K. (1982), "Hyggesstorleken under sjuttiotalet" (The Size of Clearcuttings during the Seventies), Sveriges Skogsvårdsförbunds Tidskrift 6, pp.35-45. (in Swedish with English summary)

Eckerberg, K. (1984), Naturvård på hyggen (Environmental Protection with Clearfellings), Skogsfakta, Flora, Fauna, Miljö 6, Swedish University of Agricultural Sciences, Uppsala. (in Swedish)

Eckerberg, K. (1985), "Ej godkänt för skogsbrukets naturvård - så här kan vi bli bättre!" (Non-Compliance to Environmental Considerations in Forestry: This is How We May Improve!), Skogen 2, pp.52-5. (in Swedish)

Eckerberg, K. (1986), Tillämpningen av skogsvårdslagens § 21: Slutrapport från fältundersökning (Environmental Protection with Clearfellings in Practice: Results from an Investigation in Swedish Forests), Swedish Univ. of Agricultural Sciences, Dept of Forest Economics Report 65, Umeå. (Licentiate dissertation in Swedish with English summary)

Eckerberg, K. (1987), Environmental Protection in Swedish Forestry. A Study of the Implementation Process. Department of Political Science, University of Umeå. Research Report 1987:12. (PhD dissertation)

Eckhoff, T. (1983), Statens styringsmuligheter, saerlig i ressurs- og miljöspörsmål (State Steering Possibilities, Especially with Regard to Resource and Environmental Issues, Tanum-Norli, Oslo. (in Norwegian)

Ehnström, B. and Waldén, H.W. (1986), Faunavård i skogsbruket: Den lägre faunan (Fauna Protection in Forestry: Invertebrates), National Board of Forestry, Jönköping. (in Swedish)

Ellefson, P.V. (1984), "Forest Practice Laws as a Regulatory Force: Organizational and Administrative Experiences in the United States", in Policy Analysis for Forestry Development, Proc:s of the International Conference in Thessaloniki, Greece, Aug. 27-31, 1984. IUFRO, Div.4, Vol. II, pp.21-37.

Elmore, R.F. (1979), "Backward Mapping: Implementation Research and Policy Decisions", Political Science Quarterly 94:4, pp.601-16.

Elmore, R.F. (1985), "Forward and Backward Mapping: Reversible Logic in the Analysis of Public Policy", in Hanf, K. and Toonen, T.A.J.(eds.) Policy Implementation in Federal and Unitary Systems, Questions of Analysis and Design. NATO ASI Series, Martinus Nijhoff Publishers, pp. 33-70.

Elmore, R.F, Gustafsson, G. and Hargrove, E. (1986), "Comparing Implementation Processes in Sweden and the United States", Scandinavian Political Studies 9:3, pp. 209-33.

Ericsson, L. (1985), Ett surt regn kommer att falla: Naturen, myndigheterna och allmänheten (A Hard Rain's Gonna Fall: Nature, the Authorities and the Public), Bokbox Förlag, Lund. (PhD dissertation in Swedish with English summary)

Etzioni, A. (1975), A Comparative Analysis of Complex Organizations. 2nd ed. Free Press, N.Y.

Falk, B. and Mörtnäs, A. (1984), "Multiple Use in Swedish Forest Policy", in Saastamoinen, O. et al. (eds.) Multiple-Use Forestry in the Scandinavian Countries, Commun Inst For Fenn 120, Helsinki, pp.19-22.

Goggin, M.L. (1986), "The 'Too Few Cases/Too Many Variables' Problem in Implementation Research", Western Political Quarterly 39:2, pp.328-47.

Green, P.F. (1982), Government Regulation in the Forests: Impacts of the 1973 California Forest Practice Act, Environmental Quality Series 36, Institute of Governmental Affairs and Institute of Ecology, Univ of California, Davis, Ca.

Gustafsson, G. (1987), Decentralisering av politisk makt (Decentralization of Political Power), Carlsson Bokförlag, Helsingborg. (in Swedish)

Gustafsson, G. and Richardson, J.J. (1984), "Sweden", in Ridley, F.F.(ed.) Policies and Politics in Western Europe. Croom Helm Ltd, Beckenham, Kent, pp.155-70.

Gustafsson, G. and Richardson, J.J. (1979), "Concepts of Rationality and the Policy Process", European Journal of Political Research 7, pp. 415-36.

Haigh, J.A. and Krutilla, J.V. (1980), "Clarifying Policy Directives: The Case of National Forest Management". Policy Analysis: 4, pp.409-39.

Ham, C. and Hill, M. (1984), The Policy Process in the Modern Capitalist State, Wheatsheaf Books, Brighton, Sussex.

Henly, R.K. and Ellefsen, P.V. (1986), State Forest Practice in the U.S.: Administration, Cost and Accomplishment. Station Bulletin AD-SB-3011, Agricultural Experiment Station, Univ of Minnesota, St Paul, Minn.

Hermerén, G. (1984), "Kausalanalys" (Causal Analysis), in Nilstun, T. and Hermerén, G. (eds.) Utvärderingsforskning: Kausalanalys (Evaluation Research: Causal Analysis), Muraprojektet, Report 4, Studentlitteratur, Lund, pp.8-34. (in Swedish)

Hernes, G. (1978), Forhandlingsokonomi og blandnings-administrasjon (Bargaining-Economy and Mixed-Administration, Universitetsförlaget, Oslo. (in Norwegian)

Hjern, B. and Porter, D.O. (1981), "Implementation Structures: A New Unit of Administrative Analysis", Organization Studies 2:3, pp.211-27.

Hjern, B. and Hull, C. (1982a), "Implementation Research as Empirical Constitutionalism", European Journal of Political Research 10, pp.105-15.

Hjern, B. and Hull, C. (1982b), "Helping Small Firms Grow: An Implementation Analysis of Small Firm Assistance Structures", European Journal of Political Research 10, pp.187-198.

Hull, C. and Hjern, B. (1987), Helping Small Firms Grow: An Implementation Approach, Croom Helm Ltd, London.

Hultman, S.G. (1983), Allmänhetens bedömning av skogsmiljöers lämplighet för friluftsliv, 2. En rikstäckande enkät (Public Judgment of Forest Environments as Recreation Areas, 2. A National Survey), Swedish University of Agricultural Sciences, Section of Environmental Forestry, Report 28, Uppsala. (PhD dissertation in Swedish with English summary)

Hultman, S.G. (1984), "Multiple-Use Forestry in Sweden: Undefined and Non-Existant?", in Saastamoinen, O. et al., (eds.) Multiple-Use Forestry in the Scandinavian Countries, Commun Inst For Fenn 120, Helsinki, pp.27-31.

Hydén, H. (1984), Ram eller lag? Om ramlagstiftning och samhällsorganisation (Frame or Law? On the Issues of Frame-Laws and Societal Organisation), Ds C 1984:12, Liber, Stockholm. (in Swedish)

Ingelög, T. (1981), Floravård i skogsbruket: Allmän del (Flora Protection in Forestry: General Introduction), National Board of Forestry, Jönköping.

Ingelög, T., Thor, G. and Gustafsson, L. (1984), Floravård i skogsbruket: Artdel (Flora Protection in Forestry: Description of Species), National Board of Forestry, Jönköping.

Kaufman, H. (1960), The Forest Ranger: A Study in Administrative Behavior, Johns Hopkins Press, Baltimore.

Kelman, S. (1981), Regulating America, Regulating Sweden: A Comparative Study of Occupational Safety and Health, MIT Press, Cambridge, Mass.

Kidder, L.H. and Stewart, M.V. (1975), The Psychology of Intergroup Relations: Conflict and Consciousness, Mc Graw Hill, Inc., U.S.

Knoepfel, P. and Weidner, H. (1982), "A Conceptual Framework for Studying Implementation", in Downing, P. and Hanf, K. (eds.) The Implementation of Pollution Control Programs, Policy Sciences Program, Tallahassee.

Kolam, K. (1987), Lokala organ i Norden 1968-1986: Från idé till verklighet (Neighbourhood Councils in the Nordic Countries 1968-1986: From Idea to Reality), Department of Political Science, University of Umeå, Research Report 1987:7, Umeå. (PhD dissertation in Swedish with English summary)

Lipsky, M. (1980), Street-Level Bureaucracy: Dilemmas of the Individual in Public Services, Rusell Sage Foundation, N.Y.

Lundberg, L. (1982), Från lag till arbetsmiljö (From Law to Work Environment), Liber förlag, Stockholm. (PhD dissertation in Swedish with English summary)

Lundqvist, L.J. (1979), "Environmental Impact Assessment in Sweden: Status, Problems and Proposals for Change", Policy and Politics, 7:3, pp.245-68.

Lundqvist, L.J. (1980), The Hare and the Tortoise: Clear Air Policies in the United States and Sweden, Univ. of Michigan Press, Ann Arbor, Mich.

Lönnstedt, L. (1974), En gruppering av de privata skogsägarna (A Categorization of Private Forest Owners), Liber, Stockholm. (in Swedish)

Maddala, G.S. (1983), Limited-Dependent and Qualitative Variables in Econometrics, Economic Society Monographs in Quantitative Economics, Cambridge Univ. Press, N.Y.

Mazmanian, D. and Sabatier, P. (1983), Implementation and Public Policy, Scott, Foresman and Company, Glenview, Ill.

Monteith, D.B. (1978), An Evaluation as to Voluntary Compliance with Timber Harvesting Guidelines in the Adirondacks, State Univ. of New York, College of Environmental Science and Forestry, AFRI Research Report 36, Syracuse, N.Y.

Municio, I. (1987), Från lag till bruk: Hemspråksreformens genomförande (From Legislation to Implementation: The Home Language Reform in Sweden), Department of Political Science, University of Stockholm, Stockholm Studies in Politics, 31, Stockholm. (PhD dissertation in Swedish with English summary)

Murphy, J.T. (1980), Getting the Facts: A Fieldwork Guide for Evaluators and Policy Analysts, Goodyear Publishing Co, Santa Monica, Ca.

Olsson, D. (1987), Råd och anvisningar om naturvård: en uppföljning av skogsvårdsstyrelsernas tillämpning av SVL § 21 (Instructions and Advice on Environmental Protection: A Follow-up of the County Forestry Agencies' Implementation of the Forestry Act § 21), Swedish Univ. of Agricultural Sciences, Dept of Forest Economics, Working Report 66, Umeå. (in Swedish)

Olsson, R. (1985), Levande skog: Skogsbruket i naturvårdsperspektiv (Living Forest: Forestry in an Environmental Protection Perspective), Svenska Naturskyddsföreningen (Swedish Society for the Conservation of Nature), Stockholm. (in Swedish)

O'Toole, L.J. (1986), "Policy Recommendations for Multi-Actor Implementation: An Assessment of the Field", Journal of Public Policy 6:2, pp.181-210.

O'Toole, L.J. and Montjoy, R.S. (1984), "Interorganizational Policy Implementation: A Theoretical Perspective", Public Administration Review 44:6, pp.491-503.

Ovegård, L. and Rydström, E. (1982), "Avsevärt intrång i pågående markanvändning: Två rättsfall" (Considerable Damage to Present Land Use: Two Court Appeals), Svensk Lantmäteritidskrift, pp.484-6. (in Swedish)

Pettersson, B. (1986), "Läget idag för flora och fauna" (Flora and Fauna Status Today in Sweden), in Skogen som natur och resurs: mark, flora, fauna (Forest as a Resource and Part of Nature: Soil, Flora, Fauna), Proc:s of Conference Dec. 3-4, 1985, Uppsala, Swedish Univ of Agricultural Sciences, Faculty of Forestry, Uppsala, pp. 19-22. (in Swedish)

Pindyck, R.S. and Rubinfield, D.L. (1976), Econometric Models and Economic Forecasts, McGraw-Hill, Tokyo.

Pressman, J.L. and Wildavsky, A. (1973), Implementation, 2nd ed. 1979, Univ of California Press, Berkeley, Ca.

Rustow, D.A. (1955), The Politics of Compromise: A Study of Parties and Cabinet Government in Sweden, Princeton Univ Press, Princeton, N.J.

Sabatier, P. and Mazmanian, D. (1979), "The Conditions of Effective Implementation", Policy Analysis 5 (Fall), pp.481-504.

Sabatier, P. and Mazmanian, D. (1980), "The Implementation of Public Policy: A Framework of Analysis", Policy Studies Journal, Special Issue.

Sabatier, P. and Mazmanian, D. (1983), "Policy Implementation", in Nagel, S. (ed.) The Encyclopedia of Policy Studies. Marcel Dekker, N.Y., pp.143-69.

Sabatier, P. (1986), "Top-Down and Bottom-Up Approaches to Implementation Research: A Critical Analysis and Suggested Synthesis", Journal of Public Policy 6:1, pp.21-48.

Self, P. (1972), Administrative Theories and Politics, George Allen & Unwin Ltd., London.

Siegel, W.C. and Cubbage, F.W. (1984), "Environmental Protection Law and Forest Management Regulation on Private Lands in the United States", in Policy Analysis for Forestry Development, Proc:s of the International Conference in Thessaloniki, Greece, 27-31 Augusti,1984, IUFRO, Div.4, Vol. II, pp.129-45.

Stjernquist, P. (1973), Laws in the Forest: A Study of Public Direction of Swedish Private Forestry, Acta Reg Soc Hum Litt Lund, LXIX, Lund.

Van Meter, D. and Van Horn, C. (1975), "The Policy Implementation Process: A Conceptual Framework", Administration and Society 6 (Feb), pp.445-88.

Weber, M. (1920), Wirtschaft und Gesellschaft. Transl 1962 by Secher, H.P. Basic Concepts in Sociology. Citadel Press, N.Y.

Westerlund, S. (1980), Naturvård och pågående markanvändning: En undersökning av naturvårdslagens ersättnings regler (Nature Conservation and Present Land Use: An Investigation of the Nature Conservancy Act Rules for Compensation, Liber förlag, Stockholm. (in Swedish)

Winter, S. (1984), "Flere ressourcer ingen garanti for bedre offentlig service" (More Resources is No Guarantee for Increased Public Services) in Gunst, J. (ed.) Debat om Offentlig Service (Debate on Public Services), Copenhagen: Kommunetryk, pp.121-33. (in Danish)

Winter, S. (1987), Implementation from Top to Bottom: Integrating Implementation Approaches, Paper presented at the Conference of the Nordic Association of Political Science, 16-19 Augusti, 1987, Copenhagen.

Wittrock, B. (1985), "Beyond Organizational Design: Contextuality and the Politcal Theory of Public Policy", in Hanf, K. and Toonen, T.A.J. (eds.) Policy Implementation in Federal and Unitary Systems: Questions of Analysis and Design. Dordrecht: Martinus Nijhoff, pp.17-28.

Yin, R. (1982), "Studying the Implementation of Public Programs", in Williams, W. et al., (eds.) Studying Implementation: Methodological and Administrative Issues, Chatham House Publishers, Chatham, N.J, pp.36-72.

APPENDIX I: Location of the investigated clearcuts

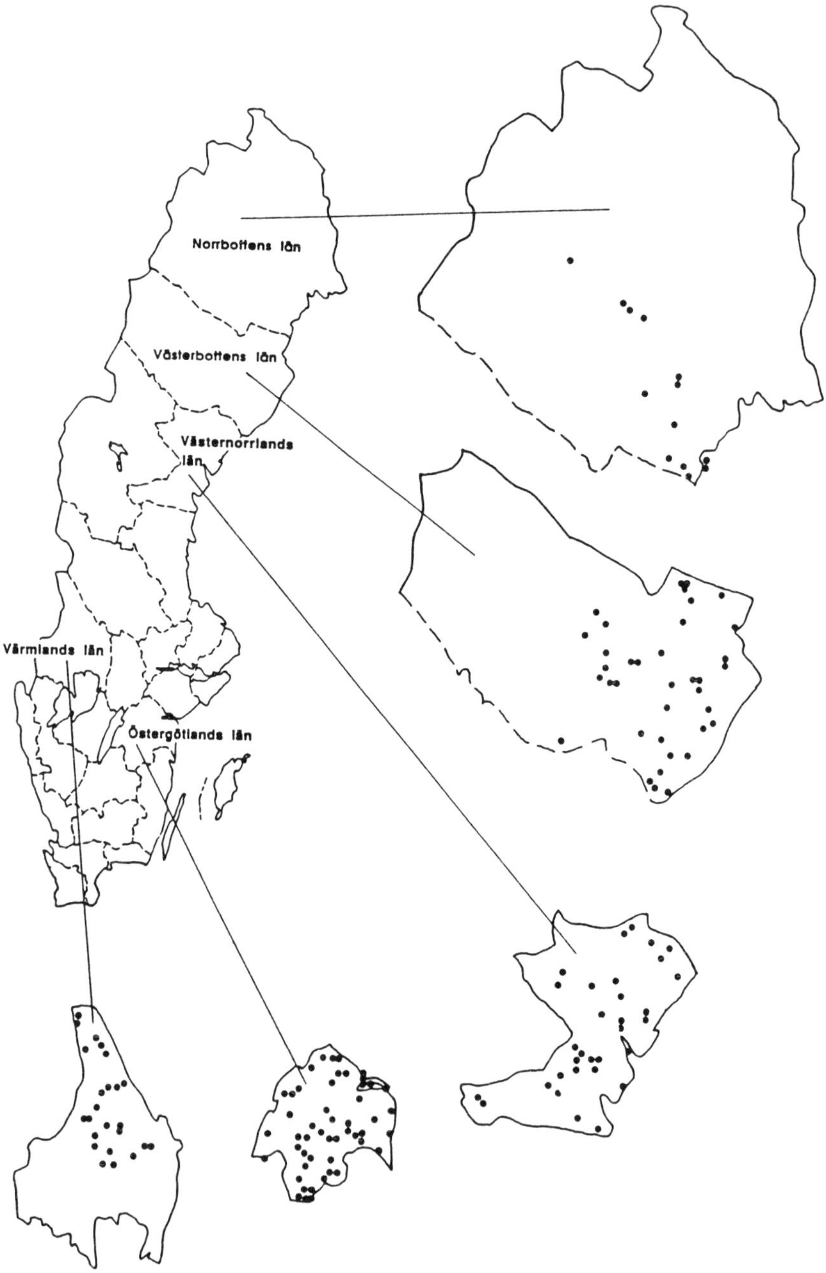

APPENDIX II: Questionnaire *

0. Introduction; presentation of the research project and of the specific clearcut investigated.

1. What people were involved with the harvest operation? Who planned and who carried out the different tasks? What were their responsibilities?

2. During the operational planning of this clearcut - what information had the planner access to - e.g. local konwledge, environmental information through maps...?

3. How long time do you estimate was used for planning environmental measures with this clearcut?

4. Were environmental features documented e.g. on a map?

5. Was any advice given to the forest workers as to what environmental protection should be considered? Who gave the advice and what was the advice?

6. What environmental features were detected on the area before harvesting?

7. What is your opinion of the result of environmental protection after harvesting?

8. Were there different opinions on what should be protected? In that case, who claimed what?

Questions 6-8 were elaborated for all the detected features.

<u>Questions 9-11 were not put to Forestry Agency officers</u>:

9. Do you know whether there is a law regulating environmental protection in forestry?

    no
    yes

    Could you describe what it implies?

10. Have you been in contact with any government representative as to environmental protection?

    no
    yes

    When and how did this happen?

---

* The questions were translated from Swedish

11. Have you, on any occasion, received a "Prescription" or an "Instruction and Advice" from the Forestry Agency as to environmental protection?

no
yes

When and how did this happen?

<u>Questions 12 and 13 were only put to Forestry Agency officers</u>

12. Have you, personally, been in contact with the Environmental Protection Section at the County Administration as to environmental protection with forest operations?

no
yes

When and how did this happen?

13. Have you, personally, goven a "Prescription" or an "Instruction and Adivce" to a forest owner as to environmental protection?

no
yes

When and how did this happen?

14. Do you have any contacts with environmentalists, e.g. members of the Society for Conservation of Nature or ornithologists?

no
yes

When and how does this happen?

15. Please describe what environmental aspects you believe are most important to consider e.g. with clear-cutting?

16. How large are "small productive areas of little importance which should be retained with forest operations if environmentally valuable?"**

** Quation from the Forestry Act regulations issued by the National Board of Forestry

17. What do you believe is "appropriate" as to limiting the size of clearcuts? Any specific limit? If so, what limit in (1) remote forest areas and what limit (2) near residential areas?

Answer: (1) .... hectares
(2) .... hectares

18. According to the legislation the demands for nature preservation should not damage forestry "to a considerable extent". The environmental measures are to be compatible with economic forestry. Have you experienced any conflicts with this - if so, how?

19. Do you remember any case where economical considerations determined the environmental protection measures? Describe!

20. From the field investigation of clear-fellings the harvest system used showed to have a large impact on the result of environmental protection. Is this in accordance with your own experiences? Give examples of how the harvest system affects protective measures!

21. Is it possible to retain nesting trees and tall stumps with mechanized felling? How? Do you usually protect these features?

22. Is it possible to avoid damage to the ground (deep tracks) with the forest machines? Are deep tracks a problem?

23. Is it possible to protect culture remnants when harvesting? Are forest professionals interested in doing so?

24. Do you believe the Forestry Act regulation has any impact on environmental protection in practice? How?

25. What about environmental protection education - would it affect forestry practice? Give examples!

26. Do you think there has been a change among the forest professionals in their attitudes and practice of environmental protection? If so, in what way has it changed?

27. Do you think there is generally enough time to plan for environmental protection measures with e.g. harvesting? If not, how should it be improved?

28. Employment

Swedish Forest Service
Private Forest Company
Forest Owners' Association
County Forestry Agency
Self-employed
Other ....

29. Are you a forest owner?

no
yes, < 50 hectares
yes, 50-100 hectares
yes, > 100 hectares

30. When are you born? Year .... .

31. What are your main duties at work?

forest worker
    manual work
    drive a machine type: ....
    other ....
timber officer, foreman
other headship position ..in forest enterprise
forest owner
    self-active in forestry
    not self-active in forestry
Forestry Agency officer
other duties ....

32. What is your basic forest education? Where?

none
secondary school, forestry line ....
forest technician ....
forest engineer
forest officer/M Sc
other ....

33. Year of exam: ....

34. Was there any education in environmental protection within these courses?

none
yes, a couple of hours
yes, a few days
yes, more than a week
do not remember

35. Have you been further educated in environmental protection after the basic education?

no
yes

What course? ....
For how long? ....
When? ....

36. What information have you seen or read among the following? (pamphlets and books were showed to the respondents)

	Not seen	Seen, but not read	Read
Forestry Agency pamphlet			
Fauna protection book			
Flora protection book			
Forestry Act regulation			
Fauna protection poster			
Fact sheets about endangered species			
Employers' instruction (if applicable)			
EPS (County administration) pamphlet (only in the county of Västerbotten)			

37. Would you like to learn more about environmental protection? How?

38. Are you a member of any society or association working with nature conservation, environmental protection, hunting etc.?